Gayle

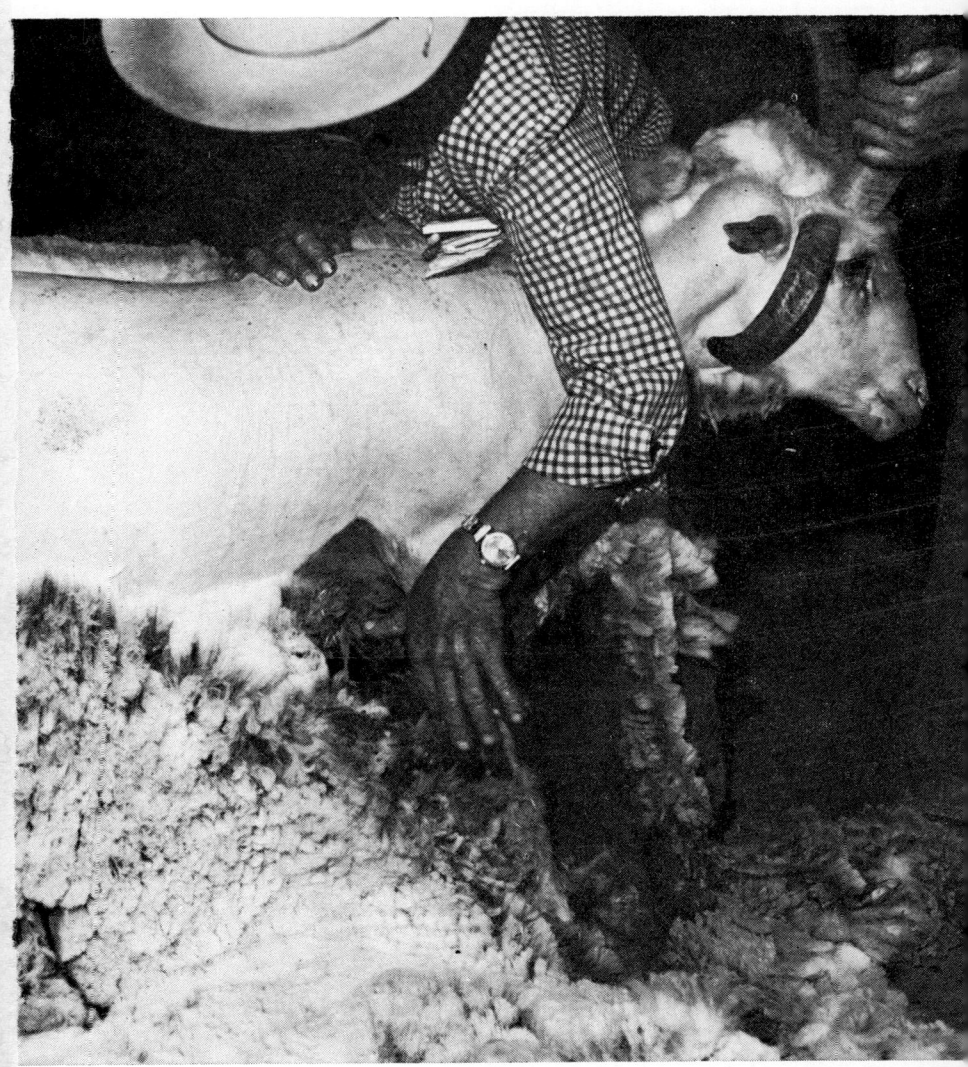

Chemical sheering of sheep. The use of cyclophosphamide or possibly other compounds in the future holds great promise for the sheep industry. This is especially true with the small breeder where shearers are scarce. The compound is introduced to the animal, and one to two weeks later the wool can be easily pulled off. The sheep illustrated here is being defleeced simply by its wool being rubbed off by hand. The skin, however, is left bare, so it must be protected from sun and weather. While not yet approved for general use, this could well become an important approved practice. (Courtesy, Dr. Glen Spurlock, University of California, Davis)

APPROVED PRACTICES IN SHEEP PRODUCTION

by

ELWOOD M. JUERGENSON, Ph.D.

Teacher Educator—Agriculture
University of California
Davis, California

THE INTERSTATE
Printers & Publishers, Inc.
Danville, Illinois

THIRD EDITION

Copyright © 1973 by

THE INTERSTATE
Printers & Publishers, Inc.

All Rights Reserved

First Edition, 1953

Second Edition, 1963

Library of Congress
Catalog Card Number: 72-83701

PRINTED IN U.S.A.

PREFACE

The **purpose** of this book is to furnish a comprehensive list of approved practices with information on how they should be done in the sheep enterprise. An approved practice in sheep production is a farm practice which has been tried and tested by State Agricultural Experiment Stations, USDA, and/or successful farmers and ranchers in the community and found to be a desirable practice to follow for most efficient sheep production. Consequently, it is important that producers know what these approved practices are and how to carry them successfully to completion in order to attain the greatest success in the sheep business.

Innovations, trends, new ideas, and research trials are also explained but not recommended unless accepted as a desirable practice by experienced sheepmen. A suggested list of approved practices is listed in the Appendix.

The **primary** aim of this book has been to carefully select and condense information so as to eliminate a vast amount of reading in order to determine the best methods and procedures to follow in carrying out the practices.

A considerable amount of information has been taken from state agricultural colleges, experiment stations, and USDA in order to provide the reader with the latest information which has proved successful. A number of plans and diagrams of sheep equipment have been included together with sources of obtaining more extensive plans should the reader desire to construct the equipment.

One of the author's chief purposes in presenting this material has been to tell **how** each activity should be done in order to accomplish the approved practices involved. Whenever possible, specific information, formulas, and amounts to use have been given. There are many activities and approved practices which are common in sheep production which, with minor local adaptations, can be used in many areas of the United States. Therefore, much of the information can readily be adapted and used throughout the country.

This book should be especially helpful to those who desire to enter the sheep industry, in addition to ranchers and farmers, full-time, part-time, and suburban; FFA members and other students of vocational agriculture; teachers of vocational agriculture; 4-H Club members; county agents; and other persons interested in efficient sheep production.

THE AUTHOR

ACKNOWLEDGMENTS

Without the kind help and assistance of many persons this publication would not have been possible. The author wishes to express his appreciation to all those who granted permission to use their plans of sheep equipment and for statements quoted from their publications.

He wishes to thank Professor J. F. Wilson, Wool Specialist, Emeritus, University of California at Davis, for his assistance in obtaining many of the photographs used in the proper handling of sheep. He is especially grateful to Robert Finley, University of California sheep herdsman, for his advice and suggestions and especially his help in obtaining photographs. Thanks also are due to Dr. Daniel Cassard, Pfizer Chemical Company, for his assistance. Special acknowledgment is given to Dr. Glen Spurlock, University of California, for his advice and help in obtaining photographs of modern practices and trends in the sheep industry. The author also wishes to thank Dr. Eric Bradford, University of California at Davis, for supplying photographs of new breeds. Appreciation is expressed to L. Heringer, a rancher of Fall River Mills, California, for suggestions on shearing ewes.

The suggestions regarding commercial sheep production by Dave Dillabo, Northern California sheepman, were valuable and appreciated. Special thanks are given to Marge and Bob Paasch, nationally famous Hampshire and Suffolk sheep breeders of Grass Valley, California, for evaluating the manuscript and supplying many exceptional pictures. The author also wishes to express his thanks to W. L. Juergenson, rancher, Auburn, California, for his help in obtaining photographs and to Mrs. E. M. Juergenson for doing many of the drawings.

Breed associations, commercial companies, the USDA, and many colleges and universities furnished the basis for most of the information. Publications which were helpful in compiling the book are from the American Institute of Cooperatives and from colleges and universities in the following states:

California	Iowa	Oregon
Colorado	Kansas	Pennsylvania
Florida	Michigan	Texas
Idaho	Ohio	Utah
Illinois	Oklahoma	Virginia

THE AUTHOR

TABLE OF CONTENTS

Chapter		Page
I.	Opportunities in the Sheep Industry	1
II.	Selecting the Breeding Stock	33
III.	Breeding and Improving Sheep	69
IV.	Handling Sheep and Lambs	93
V.	Raising Lambs	139
VI.	Feeding and Fattening Sheep	175
VII.	Shelter and Equipment for Sheep	207
VIII.	Controlling Parasites and Diseases	245
IX.	Butchering Lamb and Mutton on the Farm	291
X.	Selecting and Using Lamb and Mutton	301
XI.	Marketing Mutton, Lamb, and Wool	315
XII.	Records for the Sheep Business	347
XIII.	Producing Lamb for the Home Locker	361
XIV.	Essential Skills for the Sheepman	371

Appendix	Page
A. Associations and Publications	377
B. Tables of Weights and Measures	381
C. Summary of Approved Practices in the Sheep Enterprise	385
D. Glossary of Common Terms	393
Index	403

CHAPTER I

OPPORTUNITIES IN THE SHEEP INDUSTRY

There is an old saying among the Basque sheepmen —"Take care of the sheep and the sheep will take care of you." Such a cliché is literally true, for domestic sheep will perish if left to themselves, yet when cared for properly will return generously to their owner. The romance of the livestock industry is exemplified many times in the raising of sheep, from the nomadic tribes of the old world recorded even in biblical times to the lonely shepherd of the western plains and colorful flocks of the southwestern Indians.

Most areas of the world are suitable for raising sheep. Sheep production in one form or another ranks among the oldest of livestock enterprises, dating back to ancient times. All types of land areas including rough land, mountain, or desert are suitable environment for sheep. In recent years wool has become a secondary product compared to lamb or mutton. Nevertheless wool is a very valuable product, is a relatively non-perishable crop, capable of being stored and transported long distances to market. For this reason, sheep raising has been the vanguard of civilization and, until modern times, was known as a frontier industry. The pioneer phase of the sheep industry has passed. There are no

Fig. 1.1—Sheep can produce food and fiber on land that is too rough, too remote, or too sparsely vegetated to be used for any other kind of agriculture. Here sheep graze on such territory in a national forest as part of the multiple-use concept of our natural resources. (Courtesy, Union Pacific Railroad)

more free grazing lands. Most sheep now graze on owned or leased pasture or range on national forest.

The sheep industry is undergoing great change primarily related to its ability to market its crop. Like the days when cotton was king in pre-Civil-War southern United States, the wool industry is now gone with the wind. It is felt most keenly where the primary commodity was wool.

For years some of the wealthiest men in Australia were sheep farmers. They sold their wool at 65 cents a pound and up, and spent the proceeds as if there were no tomorrows. They tossed spectacular all-night parties in the big cities, bought fancy cars, took their wives on extravagant round-the-world trips.

Now, the once lucrative wool industry in Australia

is foundering as it is in the rest of the world, priced out of the market by cheaper man-made fabrics, the synthetics.

The average price for greasy wool fell from 44 cents a pound in 1969 to 29 cents in April, 1971.

Once-wealthy sheep farmers are working as laborers and miners, their wives and daughters as salesgirls and barmaids. Their huge sheep flocks are breaking up, and the Australian banks are foreclosing on their giant ranches, 500,000 acres and up, in the outback.

One hundred years ago 90% of the people in the United States were farmers, and 10% were non-farmers. Now the reverse is true—about 10% or less are farmers who produce an abundance of food for the rest of the people. This means that stock, equipment, and operating expenses are high so that wasteful

Fig. 1.2—An Australian sheep station. Australia has 170 million sheep. Sheep and wool are the country's economic mainstay in agriculture. This sheep station (ranch) is near Bathurst, about 120 miles west of Sydney. Sheep stations are huge in size and often located in the remote outback. It is a common practice in that country to use horses to herd sheep. (Courtesy, California Wool Growers Association)

methods mean failure; success comes only through the best scientific practices.

There are many advantages to sheep raising compared to other types of livestock. Some of these advantages are as follows:

1. Sheep produce liberally in proportion to what they consume.
2. They are one of the few farm animals that produce two crops a year. The income from these, lambs and wool, come at different times of the year.
3. Sheep are excellent weed destroyers as they eat a greater variety of plants than do other types of livestock.
4. They utilize otherwise waste land and feed on range plants not eaten readily by other animals. In addition, because of the construction of their lips they are able to pick up grain lost at harvest time in stubble fields.
5. They produce a product (wool) which can be held

Fig. 1.3—Native sheep, Liberia, Africa. They are almost woolless but produce good carcasses and are resistant to parasites and disease. Sheep are worldwide in distribution.

Fig. 1.4—Sheep make it possible to utilize rough land. Here sheep are grazing near Mt. Adams (altitude 12,307 feet) in the Gifford Pinchot National Forest, Washington. (Courtesy, U. S. Forest Service)

until market conditions are favorable. The wool crop will generally pay for the ewes' keep.
6. Sheep increase at the rate of 100% or better. Lambs also mature early so rapid return on investment can be expected.
7. They fit well into many balanced farming programs.
8. Sheep distribute their droppings well and owing to their preference for high ground, leave much of the fertilizing value where it is most needed.
9. They do not require expensive buildings and equipment or heavy labor.

Bountiful as they are, sheep are not without disadvantages. Most important is that study and continuous attention are required. Sheep are often quite helpless and fall easy prey to predators, especially dogs, coyotes, foxes, bobcats, and eagles. They might even fall prey to such hazards as picket or woven wire fences, or to ditches or gullies in which they might lie and suffocate unless aid came quickly. Parasites and disease are also ever present problems to guard against.

This pretty well summarizes the risk involved and the task ahead.

Probably the greatest disadvantage of the sheep industry is that there is little market for cull animals. In the cattle industry, for example, an old cow or aged bull will bring as much as or often more than a prime steer. This is not so in the sheep business, where only market lambs and young breeding stock bring a fair price. Old ewes and rams sell for practically nothing if they can be sold at all, yet the meat is utilized in a variety of ways. This is an area of marketing needing serious study and research in order to improve profitability in sheep.

Probably a greater variation exists in the way sheep are handled, where they are grown, and difference between breeds themselves than any other type of livestock.

Over 80% of the sheep population in the United States is found west of the Mississippi River, although the greater portion of the product is consumed in the east. In the west a large operator would be one who had 2000 ewes or more; a small sheep operator, 300 ewes or less. Commercial flocks in the east usually consist of from 30 to 50 ewes while good sized herds number from 100 to 700 or more ewes. Western sheep generally graze on native range and pasture for at least part if not all of their lives. Texas and nine mountain states account for 55% of the nation's total sheep population. Midwestern and eastern sheep generally are grown in farm flocks as part of a total farm program. Sheep can utilize vast acreages of land not suitable for cattle grazing or other agricultural enterprises. They can just as well be kept as a farm flock on com-

Fig. 1.5—Sheep fit well into diversified farming. These sheep are cleaning up in a young orchard. (Courtesy, Allis-Chalmers Company)

paratively high-priced land to utilize irrigated or non-irrigated pasture, farm grown grains, and roughage as well as to maintain fertility and provide supplemental income.

Regardless of where they are raised, sheep raising is a stable industry whether it be a small farm flock or a large sheep ranch bringing to the owners two cash crops per year.

To those contemplating entering the sheep industry, there are two factors that must be worked out satisfactorily before they begin:

1. Can you grow grass or produce other feed that is suitable for sheep economically on the land under your control?
2. Are you willing to learn the facts necessary to the proper handling of sheep?

Sheep are probably more nearly a dual-purpose animal than any other type of livestock. This tends to give

Fig. 1.6—A good flock of range ewes on native pasture. In order to enter the sheep business, feed such as this or other sources must be available at a reasonable price. Native pasture or range can often be improved by use of fertilizers, brush burning or introduction of new species of forage.

an additional variation to the numerous patterns and opportunities offered to raise sheep under different farm conditions. With the exceptions of some special conditions, the manner in which sheep are kept depends mainly upon local farm conditions, bearing in mind that an economical source of feed must be available or the enterprise will eventually fail.

The following categories represent the major types of sheep enterprises into which most producers would be placed. Those desiring to enter the industry should inventory their facilities to see in which of these divisions they might best fit.

1. Farm Flock
2. Range Producer
3. Purebred Producer
4. Commercial Feeder
5. Specialty Producer
6. Home Consumption
7. Exotics

1. Farm Flock—This generally consists of from 30

to 50 ewes. Most sheepmen consider a flock of less than 20 uneconomical from the standpoint of caretakers' time, investment in rams and equipment. It is not much more difficult to care for 50 head than 25 so that around 40 sheep or a one-ram flock is commonly found on many farms even though several hundred head on some farms would come under this classification.

These smaller flocks are more or less scavengers on most farms, cleaning up weeds, fence rows, ditches, and other feeds that might otherwise be wasted. They can often be overwintered on farm by-products such as bean straw, corn stalks, or grain stubble, which furnish the bulk of their food. Grade or crossbred ewes are used, although most people prefer a purebred ram. Whether or not early or late lambs will be produced depends upon the area under consideration. Lamb is the crop most often stressed with wool as a secondary product.

Fig. 1.7—These sheep are part of a small farm flock cleaning up a fence line near a row of trees.

Fig. 1.8—Range sheep grazing summer pasture in the Challis National Forest, Idaho. A typical range band will have from 1000 to 1200 ewes with lambs. (Courtesy, U. S. Forest Service)

Farm flocks fit in well to give a balanced type of agriculture by improving fertility, utilizing waste feeds, and, in a sense, requiring little or no labor, as the owner must be there anyway. In fact, this is so often true that the income often seems to be all profit. However, this is a dangerous point of view as many owners do not give their sheep the proper care they would a major enterprise. They also have a tendency to over-expand and get more sheep than there is waste feed, etc., available.

Nevertheless, farm flocks well cared for are profitable and will fit into many existing farm programs. Undoubtedly there are a great number of farms which could profit by keeping a proper size flock of sheep. This is one of the major areas where expansion of the sheep population of the country is possible.

2. **Range Band Method**—The romance of the sheep industry reaches its peak in the vast herds moving on unenclosed land. Whether it be on the home ranch of

a wealthy owner or following the lonely life of the Nomadic herder, there is a certain adventure in this ancient livelihood.

This is typically a western industry and the vast majority of these sheep are handled by range band method. In the southwest nearly all the range is fenced, but on other ranges sheep are herded over large areas. It is not unusual for flocks to travel over 100 miles twice each year in their annual migrations to and from summer ranges. In either case they are under the

Exterior Boundaries of Federal Lands of the Intermountain Region

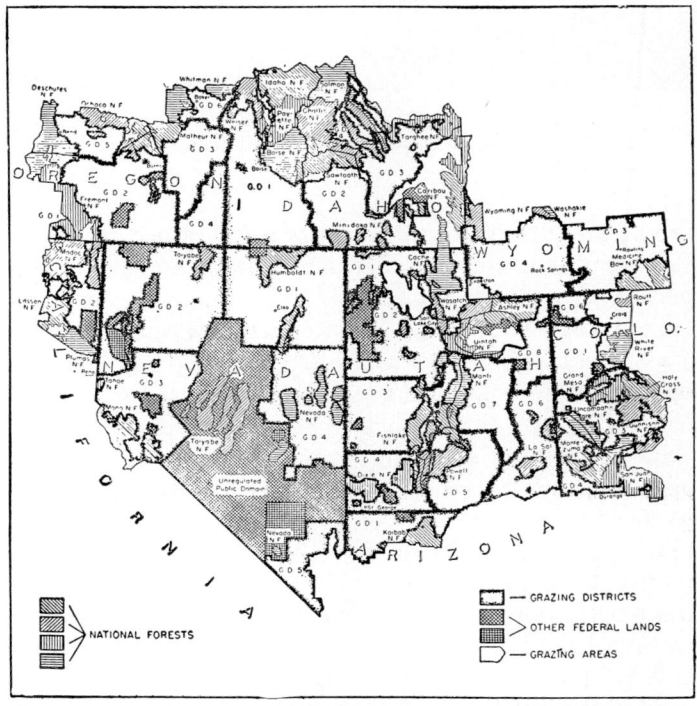

Fig. 1.9—Federally controlled lands cover a major part of this region.

Fig. 1.10—The life of the shepherd may be lonely, but it is often led in beautiful surroundings. (Courtesy, Union Pacific Railroad)

care of an expert herder and his well-trained dogs.

A definite pattern of life is followed by those who operate on western rangeland. In summer the animals graze in high mountains, most of which are in national forests. In the fall, to escape early snows, they literally pour out of the mountains down canyons to spread out on fall ranges where the ranches and private lands are located. Here the shearing, lambing, culling and other operations are carried out under the watchful eye of the owner.

Range operators with less than 800 sheep usually find it uneconomical to operate as such unless they combine with other bands. On the other hand, there are a few extremely large operators who run over 25,000 head. A typical operator would own from 2000 to 4000 head. A typical band of sheep with one herder would have 1000 to 1200 ewes with lambs, although dry sheep could run in much larger herds.

Wool is an important commodity in range operations so that most range sheep are whiteface ewes carrying 50% to 100% Rambouillet blood. This is also important because of the desirability for the sheep to band together and herd well. There is little chance nowadays to enter the range sheep industry except through inheritance or to become associated with the industry on a non-ownership basis.

3. Purebred Sheep—While many of the best purebred breeders also run large commercial herds, we will consider purebred raising as a separate enterprise. Purebred flocks vary in size, but as a general rule are smaller than range bands although at times they may be handled by range band methods.

Purebred breeders perform a useful function in devel-

Fig. 1.11—A purebred yearling Suffolk ram. It takes a long time to become successfully established in the purebred business, as it is a specialized part of the sheep industry.

oping and maintaining the best possible animals for every condition. However, the correlation between purebreds and their ultimate use and influence on commercial production must always be borne in mind. In the last analysis, the needs of commercial production will set the standard. Therefore, these breeders sell rams to commercial breeders and ewes and rams to new or established purebred breeders. These flocks must be given closer attention and one must be a much keener student of sheep husbandry if he is to be successful. Regardless of breed or location of farm, careful attention must be given to records, selection of a breed, careful selection for mating, proper feeding, etc., as well as advertising, fitting, showing, and salesmanship. One can readily see this is a highly specialized business and only those properly equipped and with a love for the business should attempt to enter it.

One of the big hurdles to overcome in the purebred business is that reputation is of paramount importance. This generally takes a long time to acquire even if it is justified and all other factors are satisfied. For this reason young people in 4-H, FFA, and other youth groups are in a more favorable position to establish themselves as purebred breeders due to the long period of time they have available to become fully established and the favorable means of publicity at their disposal. Even then it is probably a good idea for these young people to consider starting with grade ewes until they have the necessary know-how.

4. Commercial Feeders—Wherever there is a surplus of feed, either grain, roughage, or by-products, one is apt to see lambs being fattened. Commercial feeding refers to that phase of the industry where lambs are purchased and then put on feed for a short period until they are ready for market. It is common practice not to keep ewes nor raise lambs from birth but rather, to

Fig. 1.12—An excellent example of a modern market lamb. There is little wastiness, and lambs like this yield a high percentage carcass weight. This Hampshire lamb was Grand Champion of the show at the 1970 Great Western Livestock Exposition. Shown by Steve Paasch, Grass Valley, California.

purchase them. Usually western lambs, weighing 40 to 55 pounds or even up to 65 or 75 pounds, are purchased and fed until they reach a market weight of 95 to 100 pounds. Most feeders feed large numbers of lambs. For example 300 would not be considered a large number and several thousand head or more are quite common for farm conditions. A commercial lamb-feeding plant might have 20,000 or more on feed at one time. Commercial feeding falls into two types:

    a. **Feedlot fattening** is a system where lambs are

Fig. 1.13—A commercial feedlot where sheep are fattened prior to slaughter. Such a plant will fatten thousands of head a year. These feedlots are generally located near a large source of concentrate feed. (Courtesy, Allis-Chalmers Mfg. Co.)

confined in corrals and fed either hay, grain, corn, or by-products such as sugar beet pulp. This phase of feeding has all the aspects of an efficient industrial plant which utilizes machinery and all known resources to the utmost. Not many people will be able to enter the sheep industry via this route.

b. **Pasture or field** fattening is a farm operation where corn, ladino clover, winter wheat, pea stubble, or other crops grown on the farm are pastured off by lambs. This is an excellent business, as fertility is maintained on the farm and little labor and equipment are needed. Weeds and other trash are cleaned up and the ground left in good condition. Furthermore, it is an easy way to harvest a crop. Death loss may run high, as feeding is uncontrolled.

This is one method that offers a possible way to enter the sheep industry and many farmers raising

Fig. 1.14—Lambs being fattened on irrigated pasture. Some farmers make it a practice to raise pasture but own no sheep of their own and simply rent pasture on a per head or a gain in weight basis.

feed might profitably investigate feeding lambs with crops grown on their own farms. It has become a major enterprise on many of the irrigated pasture areas of the west.

Regardless of how lambs are to be fed, it can be a very speculative industry if prices are constantly going up and down. Therefore, one should study markets thoroughly and purchase only healthy, parasite-free lambs at a fair price. It will become less speculative for those who feed lambs year after year and are not continually in and out of the business.

5. **Specialty Producers**—This refers to a type of lamb production which produces a quality product more or less out of season. The Easter lamb trade, for example, prefers light weight lambs around 30 pounds and consequently must pay a premium price for such carcasses. Anyone who raises lambs for such a market is

termed a specialty producer. There are two main groups into which most producers would fall—Hothouse lambs and Spring lambs.

a. **Hothouse lambs** are produced out of season principally for eastern markets and are sold from December until April. Midwestern and eastern states produce large numbers of these lambs. Hothouse lambs are born between September 1 and January 1 and are slaughtered at 6 to 10 weeks of age when they weigh about 40 or 45 pounds. The lambs are dressed with the head and hide on and with the heart, liver, lungs, spleen, and kidneys remaining in the body. However, a good hothouse lamb will sell for as much as a 100 pound market lamb. It is necessary to start with ewes that will come in heat at any season, so for this reason the Dorset breed has long been a favorite. Of recent years crossbred ewes have become increasingly popular. The

Fig. 1.15—Dorset ewes such as these are often used to produce hothouse or off-season lambs. Hothouse lambs are born early and slaughtered at 6 to 10 weeks of age when they reach 40 to 45 pounds.

Fig. 1.16—Ewes and crossbred lambs on good pasture. These lambs were born in December and will go to market as spring lambs in April without being weaned from their mothers.

Delaine-Merino, Rambouillet, Tunis, and Dorset-Merino cross mated to Southdown bucks will yield lambs early enough for hothouse production. This phase of the sheep industry requires winter work, good barns, careful shepherding, and heavy grain feeding. Therefore, it should only be considered after one has gained experience in the sheep industry.

b. **Spring lambs** are lambs four or five months old that are marketed in the spring prior to June 1. California and southwestern states, Kentucky and the Bluegrass area, have specialized in this type of lamb production for some time. Grade ewes or crossbreds with predominantly Rambouillet blood are mated to blackface bucks (Hampshire or Suffolk) so as to drop their lambs in November and December. Such matings result in milk fat lambs ready for market in April and May weighing 85 to 100 pounds. Due to the fact that crossbred lambs are produced, it is common practice to buy replacement ewes from another area. This is a good type of production for beginners in the sheep enterprise provided they have a reasonably mild climate and good green feed for their ewes early in the spring. For this reason it is somewhat limited to those who are favorably situated geographically.

6. **Home Consumption**—Two recent developments have given considerable impetus to producing meat for the home table. One is the perfection of satisfactory home refrigeration and the other is the large number of part-time farmers who live on small farms and wish to produce enough meat for their own table. Per capita, consumption of lamb and mutton varies tremendously according to taste and location. The average for the United States is around five and one-half pounds per capita, however, some states average two to three times that much and obviously individual families would vary more.

Unless fencing is no problem and feed is available, those who desire to freeze and store lamb or mutton at home should seriously consider simply buying whole carcasses from a wholesale butcher and having it packaged into cuts specifically sized for their family use.

It is often difficult to purchase ewes about to lamb or with lambs at their side, and if only two or three ewes are kept on the farm, they are as much care as 20 or 30 and do not justify keeping one ram just for them. Nevertheless, some part-time farmers do successfully keep several ewes around the place to produce lamb for their table. Perhaps the simplest method is to buy feeder lambs and fatten them on pasture with some grain until they reach desirable weights. One or two 85 pound animals would supply enough lamb for the average family of five for a year provided regular amounts of other foods were used.

OPPORTUNITIES IN THE SHEEP INDUSTRY

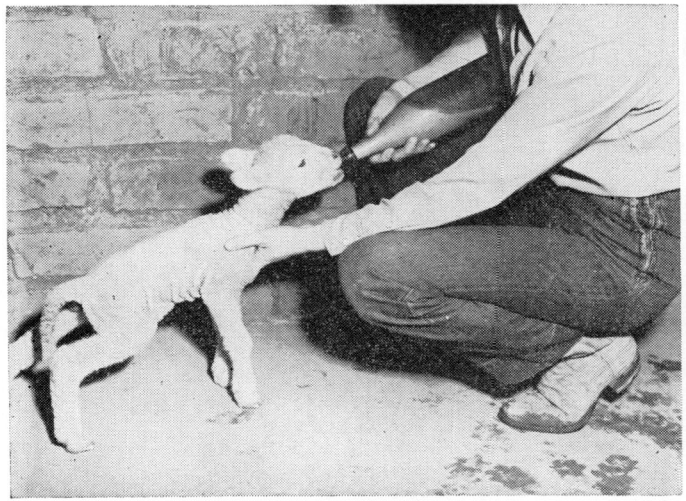

Fig. 1.17—An orphan lamb being raised at home on cow's milk. This is one good way of raising several lambs to feed out for home consumption.

Regardless of what phase of the sheep industry a person wishes to enter, he should bear in mind that sheep take constant care and that an adequate source of economical feed must be available. There have been ups and downs in all phases of the sheep industry, and there will continue to be. Nevertheless, the future of the sheep business is a stable one. The United States imports large amounts of wool, and mutton and numbers of sheep in the country are at a low level. Animal agriculture is a permanent, profitable part of our farming life. Success in sheep raising will come to those who are willing to work, study, and learn the most up-to-date methods and apply them.

7. Exotics

The introduction of foreign or "exotic" animals is not a new idea. However, as pressure mounts to main-

Fig. 1.18—Blackbuck antelope. One of many species of foreign animals introduced for hunting.

tain a profitable agriculture, all avenues of enterprise will be explored. Hunting for sport, with camera, or for aesthetic purposes suggests a definite possibility of increasing profit from rough lands, so the use of game animals will become increasingly important. Wild game, semi-wild (feral) species, or domestic crosses with wild of the same species can form the basis of an additional source of income for ranchers willing to learn and experiment in this direction. Throughout history exotics have been tried often without success, although occasionally were too well accepted. In New Zealand, for example, some 30 species have become established which have no natural predators or controls over their population. These exotics attained pest proportions, resulting in extensive damage to watershed, forest, and pastureland.

In the United States, species of European deer, African antelope, and wild boar have been well adapted to their new environment. Texas has been particularly active in establishing many species of exotics where now a source of income is derived from sale of hunting rights.

The Black-bellied Barbados, an exotic breed of sheep, is being looked at by University of California researchers. Dr. Glen Spurlock, University of California Agricultural Extension animal scientist at Davis, says "It's easy to see how these animals could become an important addition to the California scene." The colorful animal is highly fertile (twins are common, and it often breeds twice per year), and it has trophy horns and "game" characteristics, besides being extremely active and having little or no wool.

A second possibility is the Mouflon-Barbados sheep. They vary from tame to wild, depending upon how they have been treated. With a little harassment they become wild, but with feed and gentle handling become docile. The possibility exists of crossing species such

Fig. 1.19—Native goat, Liberia, Africa.

Fig. 1.20—A four-horned Navajo ram. This handsome specimen could be a forerunner of a trophy animal used as a game species. (Courtesy, Dr. Glen M. Spurlock, University of California, Davis)

OPPORTUNITIES IN THE SHEEP INDUSTRY

as these to produce game animals especially adapted to a particular area.

Multi-horned animals such as the Navajo sheep also present the possibility of trophy animals. While ranchers may anticipate additional income from exotic species from sale of trophy animals or hunting rights, numerous barriers exist. There are many ecological considerations, as well as legal restrictions—federal, state, and local. Management practices for these species, while similar to domestic production, also must be studied, examined, and incorporated in a livestock operation.

However, profit margins in agriculture and especially in the sheep industry are slim, so every possibility for profit must be examined, including production of exotic species.

Trends

The sheep industry is changing rapidly as new discoveries are made in this phase of agricultural science. These advances should result in more efficient production and eventually greater profit to the producer while at the same time producing better quality food at a reasonable price.

D. S. Bell, Department of Animal Science, Ohio State University, has this to say about the future of the sheep industry:

"In Ohio, and likely in all of the North Central States, production will be intensive. Ninety pound lean-meated lambs produced in 100 to 110 days will be an increasingly common practice. Development of ewes that will produce two or more lambs with each pregnancy will be an important objective. Diet for the lambs will be improved so that twins will gain as fast as singles after 30 days of age. Already rations and methods of feeding have brought twin lambs along nearly as rapidly as the singles can be made to grow. Heat cycle, range of breeds and types will be the subject of study, and those that show a broad range of breeding activity will gain preference so that ewes can be made to work more intensively; perhaps three crops of lambs each two years is not an unreasonable expectancy. Ewe flocks will be used to graze the grass and lambs may never be allowed out of the barn. This alone will be a big aid to parasite control since lambs have been shown to be from six to ten times as susceptible to infection as older sheep. Eventually, a year-round supply of delectable spring lambs of three to four months of age will be available. Regularity of supply will become increasingly important if lamb is to become a regular dietary inclusion for our increasing population."

Much of what is said of Ohio will be true for all sheep producing areas. In addition, other developments are occurring which will influence sheep production. The use of frozen semen in artificial insemination and the control of heat periods by drugs are meeting with success in experimental trials. Such a combination will speed the production of more than one lamb crop per year as well as improving management and feeding practices. Additional experimentation suggests merit in weaning lambs early and placing them on concentrate feed so ewes can be rebred rather than nursing lambs. Lambs raised in this fashion may even out-perform milkfed lambs.

By using a breed like the Finnish Landrace, that commonly produces multiple births—generally two to five, with an average of about three—the percentage of lamb crop may be materially increased, thereby lowering the cost of production.

While some new diseases and parasites appear others are being conquered. One state (Wisconsin) has practically eliminated scabies with an intelligent statewide attack on this disease.

Fig. 1.21—A Finnish Landrace ewe with quadruplets. Ewes of this breed tend to have litters of lambs, rather than singles or twins. They often give birth to three to five or six, although the breed average is about 2.9.

Other advances could be cited which would emphasize the bright future for the sheep industry. It is evident that success will not come with slipshod production and marketing methods. The modern producer must keep up-to-date, and the new producer must do everything possible to gain know-how as provided in this text and other up-to-date sheep publications.

A joint task force of the U.S. Department of Agriculture and the state universities and land-grant college was formed in 1968 to participate in a long-range study for a National Program of Research for Sheep and other animals. The summary of their introduction on sheep and wool is as follows:

Summary—Sheep and Wool

A. **Importance and Nature of the Sheep Industry**

   In 1968 there were 22,000,000 sheep and lambs on 200,000 farms in the United States.
   Annual sales of sheep, lamb and wool amount to 1% of the cash farm receipts from all commodities . . . lamb makes up 80% and wool 20%.
   World sheep numbers have risen from 750,000,000 head in the late 1930's to just over 1,000,000,000 head now . . . world wool production is 6 billion pounds compared with 5½ billion a decade ago . . . 10 countries have ⅔ of the sheep.

   1. **Geography of Sheep Production**

      All the states produce sheep . . . Texas is the leading State with 24% of the nation's sheep numbers . . . 80% of the sheep in the U.S. are west of the Mississippi River.

B. **Historical Trends in the Sheep and Wool Industry**

   The nation's peak in sheep numbers of 56,000,000 head was reached in 1942 and declined to 22,000,000 in 1968.

   1. **Trends in Production**

      In 1910 over half the returns from sheep production came from wool—now about 20%.
      Improvements in production have been limited. Shearing with chemicals a promising new development.
      Wool Act of 1954 provided incentives for wool production with resulting increases for a while, then a decline.

   2. **Trends in Processing and Product Development**

      Lamb slaughter has shifted from major terminals to supply areas.

Fewer plants now slaughter sheep.

Increased retailer bargaining power has led to more specification buying.

New processing and distribution methods are emerging.

Wool textile mills are shifting from the Northeastern States to the Southeast.

Manmade fibers have become increasingly serious competitors of wool; however, wool has held up quite well.

3. **Trends in Marketing**

A 1955 survey showed lamb available in only 39% of the nation's retail stores handling red meats.

Promotional efforts have shown that sales of lamb can be stimulated.

Lamb grades became an issue in 1959 leading to revisions of the grades and a research study of their effects.

Cutability grades have now been added.

Imports of lamb meat have tripled during the last decade.

Only limited improvements in wool marketing have occurred.

4. **Trends in Consumption**

Per capita consumption of lamb has declined from a peak of 7.7 pounds in 1912 to 3.7 in 1968, compared with gains for beef and chicken.

Per capita use of wool has declined from a peak of 5 pounds in 1941 to 2.3 pounds in 1968.

C. **Activities Aimed at Helping the Sheep Industry**

A 1950 study suggested a national sheep flock of 37,000,000 head for utilizing feed and forage resources especially suited to sheep.

A National Wool Act of 1954 provided production incentives and authorized a self-help program.

Promotional activities were conducted and national conferences were held.

The Wool Processing Laboratory was established at the USDA Western Regional Research Laboratory in 1959.

The McKinsey Company was engaged in 1961 to make a study of the industry.

The sheep industry developed specifications for consumer preferred lamb and procedures of preparing wool for market in 1964.

The National Meat Animal Research Center was established in Nebraska in 1964 with 25% of its program devoted to sheep.

The Sheep Industry Development Program was established in 1967 to implement more efficient production and marketing.

D. **Research Perspective and Comments**

Only recently have increased efforts been made to update the pastoral nature of the sheep industry.

In spite of many pessimistic factors pertaining to the industry there is still room for optimism.

The sheep industry has much un-used potential.

Both lamb and wool are deficit products nationally.

Land and roughage resources are available in abundant supply that would not be used except for production of sheep.

Sheep are capable of converting roughages into a high quality food protein.

Sheep raising and wool production are important in certain parts of the nation . . . they are farm enterprises that may have potential in rural poverty areas.

Lamb is not available to all persons who might wish to eat it.

Wool still remains unmatched for many textile uses
... manufacturers of competing fibers find it advantageous to blend wool with their product.

Lamb and wool have not fully satisfied consumers
... research has helped and offers more help for the future.

A 56% increase in overall research effort is recommended.

CHAPTER II

SELECTING THE BREEDING STOCK

Regardless of what phase of the sheep enterprise one is going to enter, the selection of breeding stock is a fundamental part of the business. Many of these principles are the same for small raisers as for those who plan to enter or are operating on considerable scale. Feed cannot be efficiently utilized nor the farm facilities properly coordinated with livestock unless the right kind of sheep are used as breeding stock. Therefore, it is of considerable importance that the right kind of practices be considered and used in selecting breeding stock.

Activities Which Involve Approved Practices

1. Selecting a type adaptable to your farm.
2. Selecting the proper breed.
3. Selecting feeder lambs for fattening.
4. Selecting breeding ewes.
5. Selecting breeding rams.
6. Buying at the right time.
7. Selecting for high productive capacity.
8. Deciding on number to stock.

Fig. 2.1—These animals are Angora goats. This is a good example of selecting the right breed for your farm, as this land was formerly very brushy and has been reasonably well cleared by grazing goats.

1. Selecting a Type Adaptable to Your Farm

Sheep have been molded by man to yield any product he desires, be it lamb, mutton, wool, or even pelts suitable for fur. New enterprises have been developed by getting lambs (hothouse) on the market during unusual times of the year. Therefore, beginners should thoroughly study which of these products will yield the most income for their particular farm.

Find out your best market—This may not always be the highest price. For example, per animal, raising purebreds would probably offer the highest gross financial returns, yet there are many other very important items to be considered before deciding to be a purebred raiser. In addition, it may not always mean the greatest net profit. On the other hand, good markets for hothouse lambs, spring lambs, or a good market for feeders produced in your area may all help in deciding what

SELECTING THE BREEDING STOCK

type is best suited to your conditions. In general, it is a mistake to produce something that is novel or foreign to your district simply because you will have a monopoly on the market. A good idea is to look to progressive farmers in your area and follow what they have found to be most desirable from a market standpoint.

Select according to your farm—Geographic location, climatic conditions, size of farm, type and availability of feed, and labor all enter into determining what type of sheep are best on a particular ranch. For instance, wool-type sheep are better adapted to open range conditions since they possess herding instinct, and wool and mutton are given equal emphasis. Small farms may raise either wool or mutton types, but as meat is most desired and fields are fenced, mutton breeds work well here. If one were going to raise purebreds, he must consider the time and additional labor required. For this reason purebred raising would not fit very well into a farming program where other crops are raised and sheep are of minor importance.

In spring lamb production, a crossbreeding program is a desirable method, as climatic conditions permit early lambs from wool type ewes which fare well during the winter and spring months, yet they are sired by mutton-type bucks in order to gain scale and mutton qualities in the lambs.

The following chart outlined by **Cornell Bulletin** 828 shows the relationship between the various types of sheep:

CLASSIFICATION OF SHEEP ACCORDING TO TYPE

Fine wool type	Fine wool breeds	Rambouillet Delaine-Merino American-Merino
Mutton type	Long wool breeds	Lincoln Leicester Cotswold Romney Marsh Black-faced Highland
	Medium wool breeds	Southdown Shropshire Hampshire Suffolk Oxford-Down Tunis Cheviot Dorset Targhee Romeldale Panama Corriedale Columbia
Fur type		Karakul

2. Selecting the Proper Breed

Within limits, the selection of a breed is not as important a factor in sheep production as are many other phases of sheep husbandry. There are successful breeders in all breeds providing the proper type has been selected to fit into market demands, feed, climatic conditions, or other limiting factors on your farm.

Choose according to personal likes—One is more apt to be successful in raising any type of livestock if the breed he is handling fits his own individual preference. More pride is taken in the flock and as a result, general care and attention is always at a high level. Therefore, this consideration should not be overlooked provided it is not in direct contradiction to important factors such as feed conditions or market considerations.

Select good individuals—This is especially important

with purebred breeders. However, in any case it is always better to have excellent individuals regardless of what breed they are than to have poor animals of a popular breed or a breed that suits your fancy. Good individuals, regardless of breed, will generally make a profit for their owners whereas this is not always true for poor animals even of a popular breed.

Buy from reliable breeders—This is quite important for beginners as they cannot always rely on their own judgment as an experienced sheepman would be able to do. Even those in the business for a long time deal only with reputable breeders. Local breeders or neighbors are often the best source, and many times it is a mistake to go to unfamiliar areas until local sources have been exhausted. Breed associations and county agents can often give indications of reliable breeders where one can obtain suitable stock.

Select a common breed—The average person starting in the sheep business will do better if he starts with a recognized, well known breed rather than a little known breed whose sponsors promise unusual qualities for the breed. As a general rule, it is desirable to choose a breed common to your local community.

Breeds of Sheep

While there are several hundred breeds of sheep in the world, there are only a dozen or so that are important commercial breeds in the United States.

The better known breeds of sheep may be grouped into two types—the **fine wool** and the **mutton** type. The mutton type may be further subdivided into medium wool and long wool. Some authorities would also list the **carpet wool** type such as the Black-faced Highland breed, and the **fur** type.

Fine wool type—These breeds have individual wool

fibers much finer and smaller in diameter than do other breeds. Annual yield of grease wool to the sheep, with few exceptions, is much greater than breeds of mutton type. The American-Merino (A & B types) are decidedly

Fig. 2.2—Six head of excellent type, long wool, mutton sheep. These Shearling (yearling) Romney rams are from Ashford, England.

lacking in mutton conformation and are of little economic importance today. On the other hand, the smooth bodied C type Merino and the Rambouillet have better mutton qualities and are popular over a wider area.

Mutton type—Sheep of this type have been bred and selected for their ability to produce lamb and mutton economically. Wool production, while not equal to fine wool types, has not been entirely overlooked because it is still of considerable value. The **long wool** breeds produce much longer fleeces, the individual fibers of which are larger in cross section than those found in the **medium** wool breeds. The Corriedale, Columbia, Panama, and Romeldale are listed as medium wool breeds but could also be listed as crossbred-wool-type sheep.

Fur type—The fur type is represented by the Kara-

SELECTING THE BREEDING STOCK

kul breed from which the pelts from very young lambs are sold for fur garments. Older lambs of this breed may be slaughtered for food and the wool from the breeding flock used by manufacturers.

Characteristics of Common Breeds

Some of the characteristics of the breeds, according to the breed associations, are as follows:

The **Cheviots** are a hardy, active breed of good quality. Cheviots take their name from the Cheviot Hills of Scotland where the breed originated several centuries ago, and gave their name to the famous Cheviot tweeds.

As early as the year 1372 it is recorded that there was a very hardy, white-faced race of sheep roaming over the Cheviot Hills, along the border between England and Scotland. These sheep were eager and able to seek out their feed over large areas of the rugged

Fig. 2.3—A desirable type Cheviot ewe. Cheviot ewes lamb easily and are excellent mothers.

hillsides, and to use it to good advantage. Being alert and swift, they were quick to sense danger and could escape many of their natural enemies. They dropped their lambs with ease, and gave an ample supply of milk to make their lambs grow fast. They had wool on their bodies where wool grows clean and long, with none on their legs and head where wool grows short and is difficult to shear.

The face and legs of the Cheviots are bare of wool and covered with short, white hairs. They are small in size, light-fleeced, and do not herd well. They produce lambs that finish well. Fleece weight is from five to seven pounds. Mature rams weigh 160 to 200 pounds, and ewes weigh 120 to 160 pounds.

The **Columbia** breed originated in America. Both sexes are polled, open-faced, and possess good herding instinct. Ewes produce a clip of 11 to 13 pounds, grading three-eighths to one-fourth blood. Mature rams weigh 225 to 275 pounds or even more, and ewes generally weigh 125 to 190 pounds. It is readily seen, then, that it is a large breed and at present more popular in the west, although some are found in the central states. In order to register animals, the association requires inspection of animals before being admitted. This is the only sheep registry association with this stipulation. Registration is increasing and the breed is growing in popularity.

The **Corriedale** is one of the crossbred wool breeds, and has been known for a long time. It is a medium-sized breed with white color markings and relatively open-faced. They are good wool producers and are gaining in popularity. They were developed in the United States by crossing long-wooled breeds with Merinos or Rambouillets. Both sexes are polled. They shear from 10 to 12 pounds under range conditions which characteristically grades a three-eighths blood.

SELECTING THE BREEDING STOCK

Fig. 2.4—A good type Corriedale ewe. Note the relatively open face and rugged appearance.

Size: Inasmuch as mutton is marketed by the pound, size in Corriedales is of the first importance. Ram lambs should wean from their mothers at or above 90 pounds; ewe lambs at somewhat less. Yearling rams at 18 months should weigh 175 pounds, or more, and the stud ram capable of siring offspring in this weight achievement range must mature in field condition at the 275 pound mark. Mature ewes should weigh 165 pounds and up.

The **Dorset** differs from other medium wool breeds in that the ewes breed any time and thus can produce off-season lambs. They are light wool producers, and both sexes are horned. The face, legs, and hoofs are

Fig. 2.5—Dorset breeding ewes. These are active, alert sheep, that produce good lambs. They are noted for breeding at any time of the year, so off-season (hothouse) lambs are a specialty. (Courtesy, Continental Dorset Club)

free from wool and white in color. Mature rams weigh 175 to 250 pounds, and ewes weigh from 125 to 175 pounds. Dorsets do not rank high in wool production, shearing about seven and one-half pounds of combing wool, grading three-eighths to quarter-blood or 48's to 56's. The ewes produce more milk than other breeds, and birth of twins and triplets is common. Dorset ewes are the earliest breeders of all and will breed in the spring of the year to lamb in the fall; this makes them the most popular breed among the producers of hothouse lambs.

The **Hampshire** is a very popular, well known mutton breed. They shear 6 to 12 pounds which is lighter than Shropshires, but produce larger, faster growing lambs. For this reason, they are used extensively for cross-

SELECTING THE BREEDING STOCK 43

breeding purposes, and in the production of spring lambs. Both sexes are hornless. Face, ears, and legs are black with the ears tending to stick straight out from their heads. They do not herd as well as fine wool breeds and, therefore, are often found in farm flocks. Rams weigh 225 to 300 pounds, and ewes weigh 150 to 200 pounds.

The Hampshire breed was founded in Southern England, mainly in the country of Hampshire. The foundation stock that went in to make up the Hampshire were

Fig. 2.6—First place yearling ewe and Reserve Champion ewe at the 1970 Western National Hampshire Show at Portland, Oregon. Notice the clean lines, strong legs, and open face. This breed is a large, well known mutton breed used extensively for crossbreeding purposes. (Courtesy, Paasch Ranch, Grass Valley, California)

native sheep known as Wiltshire Horne and the Berkshire Knot. These two breeds were crossed with Southdowns. The Wiltshire Horne and Berkshire Knot were each a large breed of sheep, hearty and prolific, but a little tall and lacking somewhat in mutton tendency; introducing the Southdown cross improved the mutton quality of the breed which we now know as the Hampshire-Dam. The Hampshire breed has been kept pure for over a hundred years without the use of any outside blood.

One of the most exciting breeds new to America is the Finnish Landrace. The important feature of the ewes is their ability to have multiple births. They are known for having litters of lambs rather than singles. Litters of three are most common and they may have up to six or seven with an average around 2.9.

The Finnish Landrace sheep are of the group generally referred to as the Scandinavian Shorttails. The breed was developed in Finland and according to the breed literature the sheep have been bred for scores of generations with no intermixture of other blood. The Finnish Sheep Breeders Association—the Lampaanjalostusyhdistys—dates only from 1918.

In discussing the breed Dr. Glen Spurlock of the University of California, Davis, comments:

"The whiteface breed is relatively small in size with fine bone, small ears and slender neck and muzzle. The tail is short. Crosses with larger U.S. breeds can be expected to be larger and with higher growth rate. Limited experience with the Landrace rams and with young crossbred lambs at Davis indicates that the Landrace is very gentle and appears to enjoy contact with humans. The phenomenon has been reported elsewhere."

Healthy ewes apparently weigh from 120 to 140 pounds at maturity, although some would weigh more if feed conditions were right. Rams range from 175 to

SELECTING THE BREEDING STOCK

Fig. 2.7—A Finnish Landrace ewe who has given birth to four lambs. Generally only two are kept and the extras sold. This breed holds great promise for crossing with other breeds in order to greatly increase the percentage of lamb crop. (Courtesy, Dr. Eric Bradford, University of California, Davis)

200 pounds, with a few individuals slightly heavier. Wool is soft, lustrous, elastic, with kemp fibers rare, and varies in fineness from 48's to 58's, with scouring loss around 30%. The average ewe produces about 5.5 pounds annually.

The important factor, of course, is their multiple-birth characteristic, which hopefully can be introduced into other breeds in an effort to increase efficiency of lamb production. While many sheepmen prefer no more than two lambs per ewe, there is a ready market for additional lambs, and multiple births can raise the flock average. Dr. Spurlock indicates:

"If you could produce litters of lambs, you could wean the extras at a day or two of age—after first getting some colostrum from the ewe—and start them

Fig. 2.8—A Merino ram. This breed is very old historically. Many present day breeds trace part of their ancestry to the Merino breed. They are noted for fine wool production.

on cold artificial milk. Then you could advertise for people to come, bring the family and buy a 4-H or FFA project lamb or just a pet."

The **Delaine Merino,** or C Type, is a somewhat small sheep, but has fine wool and is noted for its flocking instinct. Ewes are hornless and color markings are the same as for Rambouillet. Historically, it has contributed more to the sheep industry than any other breed, primarily because of a strong herding instinct and its fine wool.

No breed can boast a longer line of royal ancestry. Its history traces back into the realms of legend and tradition. Even in the days of the Caesars, the golden fleece and the fabrics made from the Merinos were the staple articles of commerce in the then known world.

The **Oxford** is the largest of the medium wool breeds and a heavy wool producer. There is a moderate extension of wool on the face and legs, and a marked tuft of wool on the forehead. Color markings range from

SELECTING THE BREEDING STOCK

brownish-grey to dark brown. There is some complaint that lambs of this breed do not fatten easily until they are too heavy for top quality carcasses.

Oxford-Down Sheep had their origin in England well over a century ago. About 1830 some English sheepmen decided there was need for a certain type and size of sheep with the productive capacities that were sought at that time. Breeders found these qualities in the sheep now known as the Oxford-Down. The name Oxford, or Oxford-Down, was chosen because early development of the breed took place in the Downs section of England and, more specifically, in the shire, or county, known as Oxford.

The Oxford sheep originated from selected individuals which were of the long wool type, and these were mated with the best medium wool sheep from the county of Hampshire.

There was sufficient progress in developing a uniform type so that by 1840 attempts were made to get Oxfords recognized in the leading shows. In 1862 Oxfords were regarded as a breed, and a class was provided for them at the Royal Society Show, as a distinctive and true-bred breed.

The first Oxford importations were made in 1946 by Clayton Reybold of Delaware. The American Oxford-Down Record Association was organized in 1882 at Xenia, Ohio.

The **Lincoln** is the largest of the long wool breeds. Color markings are white, and both sexes are polled. While light in yolk, the 10 to 14 inch staple may weigh 15 to 20 pounds per year's growth. Fleeces are heavy with long, coarse, strong, and lustrous wool. The wool is used primarily in carpet manufacturing. Because mature rams weigh 250 to 350 pounds and ewes weigh 225 to 250 pounds, the Lincoln is possibly the heaviest breed known.

Fig. 2.9—One of the largest sheep of the long wool breeds is the Lincoln. Note the long, lustrous wool on the above animal.

The **Rambouillet** is especially popular in the range sheep country, and forms the basis of the industry. The breed is a heavy producer of wool. Some skin wrinkles exist in the Rambouillet breed, but the C type (smooth body) is by far the most popular. As a general rule, the rams have horns. These sheep herd well, are prolific, and raise good lambs, although crossbred lambs from Rambouillet ewes and Suffolk or Hampshire bucks are much superior in weight gains and quality of carcass. Mature rams weigh 225 to 275 pounds, and ewes weigh 140 to 200 pounds.

The foundation sheep breed of the American sheep industry originated in France. Parent stock was from the Spanish Merino, and no other blood has ever been introduced. In 1785 Louis XVI, impressed by the importance of wool and wool manufacturers in the industrial growth of France, asked the King of Spain, as a per-

SELECTING THE BREEDING STOCK

sonal favor, for "permission to import from the celebrated Spanish flocks a flock of sheep with the highest quality wool." His request was granted and in October, 1786, 318 ewes and 41 rams, representing the best that French Agent Gilber could find, were quartered in their new home on the government farm at Rambouillet, near Paris. From 1786 to the present time, the carefully kept records of the French flock have been preserved without a break.

Rambouillets were brought to the United States in 1840, but at that time the American-Merino was coming to the fore and the French sheep did not get a favorable reception in the east. When California began to be a place of importance after the Gold Rush of 1849, these

Fig. 2.10—A Rambouillet ewe. These sheep are excellent wool producers, large, rugged, and popularly known as white-face sheep in the range country. Rams are horned.

French sheep were gathered up and sent to the Pacific Coast where they served as the foundation stock of the California French-Merino.

Although a few breeders in Ohio and Michigan bred Rambouillets in a rather quiet way, it was a German, Baron Von Homeyer, who introduced the Rambouillet as such to the United States and attracted the attention of the sheepmen of this country to them. He exhibited his Rambouillets at the Columbia Exposition in Chicago in 1893 through his American Agent W. G. Markham of

Fig. 2.11—Champion Southdown ewe. While somewhat smaller than most breeds, they are extremely blocky and yield excellent carcasses. Some showmen like to cross them with larger mutton breeds. (Courtesy, American Southdown Breeders Assn.)

Avon, New York. Von Homeyer's sheep were so exceptional in size that people gazed on them in wonder. To the breeders of American-Merinos, they seemed an almost impossible creation out of Merino blood.

The **Southdown** is a medium to small-sized breed of sheep and often referred to as the ideal mutton type. The body, which is oval on top, is wide, deep, low set and evenly covered with deep, firm flesh. The lambs are often fat at lighter weights than any of the larger breeds. The weight of fleece ranges from five to eight pounds and in twelve months it attains a length of about two inches of the very highest quality. Mature rams in good flesh should weigh from 175 to 200 pounds and ewes should weigh from 125 to 150 pounds. Both sexes are hornless. In crossbreeding and in grading up, Southdown rams are extremely prepotent and used extensively for this purpose. The Southdowns have been successful in producing grand champion wethers, truck loads and in carcass competition, and are very popular for Junior Club work in 4-H and FFA. They are one of the Down breeds originating in the British Isles.

The **Shropshires** have been called "the farm flock favorite" and the "middle-of-the-road" type. They are quickly recognized by the downy, white wool covering most of the face and legs, and from the dark brown, buttonlike nose set between two puffy jaws. They have short, alert ears which stick almost straight out from the sides of the head, a broad chest, thick loins and deep, bulging hindquarters. They are considerably larger than the Southdown but smaller than the Hampshire. Mature ewes weigh 150 to 180 pounds, and rams weigh from 180 to 250 pounds.

The Shropshire breed has good mutton qualities and is one of the heaviest wool producers among the medium wool breeds, shearing an average of 8 to 12 pounds. Twinning is high, often averaging 125 to 175 lambs per 100 ewes. The mothers are gentle, easy to manage, and are good milkers. The lambs develop rapidly with a minimum of grain feeding, and on the market the Shropshire carcasses are pushing Southdowns for the money.

The main criticism of the breed is "wool blindness." Not long ago breeders deliberately bred wool onto the face to give their "Shrops" a "natty" look. It was soon realized that the wooly heads were not practical and the more progressive Shropshire breeders are now breeding to correct this excessive face covering.

This breed of sheep originated in the counties of Shropshire and Staffordshire, in central western England. Records are not clear as to how the breed was developed. Some maintain that it was formed by selectting and mating the best from the old native breeds of the two counties, while others say that it came into

Fig. 2.12—Typical Shropshire ram. At present this breed is very popular as a farm flock. (Courtesy, American Shropshire Association)

Fig. 2.13—Modern Suffolk sheep. Pictured here are some of the best animals representing this breed at the 1970 Cow Palace in San Francisco. Note how those showing sheep are in control of the animal, yet down and out of the way, with always an eye on the judge.

existence through the crossing of improved Southdowns, Leicesters and Cotswolds and the native blackfaced, horned sheep that were known as Longmynd. Southdown rams were used to breed out the coarseness and horns of the original sheep while Leicester and Cotswold blood improved the length of the wool and gave size to the Shropshires. When these various elements had become sufficiently fused, the breed became known (1848) by the name it now bears.

The rise of the breed was rapid. Shropshires were first exhibited in a special class for short-wooled sheep at the Royal Show at Gloucester in 1853. In 1859 they were classed as a separate breed.

The first importation of Shropshires into the United States was made into Virginia about 1855, although there is no definite record of who imported these sheep

Fig. 2.14—A fine Suffolk ewe. The open face and clean legs make these sheep popular for crossbreeding on western ranges. They produce large, mutton-type lambs.

or the disposition that was made of them. In 1860 a ram and 20 ewes were brought to Maryland by Samuel Sutton, and by 1880 most of the farm flock states were liberally supplied.

The **Suffolks** are large and upstanding. While poor wool producers, they are good mothers and yield large, fast growing lambs. Crossbred producers, using Suffolk bucks, rate them excellent for this purpose, and they are gaining in popularity. This is especially true on western range country and for spring lamb production. Color markings are black with no wool on the head and legs. This is a desirable factor for commercial producers. Both sexes are polled. Mature rams weigh 225 to 300 pounds, and ewes weigh from 160 to 225 pounds.

The Suffolk breed originated by crossing Southdown rams with Norfolk Horn ewes. The Suffolk gets its

SELECTING THE BREEDING STOCK

quality of mutton and wool from the Southdown and its hardiness and productiveness from the Norfolk. The breed is of English origin and has reached its high standard of perfection through the interests of the English Suffolk Society. Since England is a country with a special reputation for its mutton, the chief object of English shepherds is to breed sheep which most nearly conform to a choice carcass.

The **Targhee** is a newcomer to the sheep industry compared to the English Down breeds. It was developed primarily for use in the intermountain area of western United States by the United States Department of Agriculture Experiment Stations at Dubois, Idaho. Interbreeding has been carried on since 1926, using primarily the Rambouillet breed. Crosses have been made using Rambouillet rams on Lincoln-Rambouillet

Fig. 2.15—A desirable type Targhee ram. The above animal is a two year old stud ram breed by the Mount Haggin Livestock Company and owned by Ohio Agricultural Experiment Station, Wooster, Ohio. At the time this picture was taken he weighed 267 pounds. Note the open face and strong upstanding body.

and Lincoln-Rambouillet-Corriedale ewes. The Targhee is clean-faced, free from wrinkles, and polled. Mature rams weigh about 200 pounds and ewes weigh about 120 pounds. Ewes shear about 11 pounds annually of 3 inch staple half-blood quality wool.

3. Selecting Feeder Lambs for Fattening

It is always a problem to find thrifty lambs of the right weight to feed. Like any other feeding enterprise, lambs bought right are half sold. Good feeders will show a profit, whereas the wrong kind may not only fail to make money for the owner, but lose money as well.

Be sure of your feed—Profit will result in feeding only if good feed is available so that lambs will gain in weight. Therefore, before lambs are purchased, a sure supply of good feed must be on hand. Lambs cannot merely be maintained as ewes might be in the anticipation of another crop of lambs. Many different feeds will do a good job of fattening. Grain or other concentrates are excellent feeds to fatten them on. Good pasture, especially when predominantly legumes, is excellent, or a dependable source of some by-product like sugar beet pulp will fatten lambs readily.

Buy at right age—Lambs too young may shrink too much, especially when taken away from their mothers. Four months of age is all right, but six months old is probably better. Lambs weighing from 55 to 65 pounds are the most desirable weights to purchase for feeders. Occasionally one may have to take them heavier.

Feed quality lambs—Here is a situation where you can disregard breed as any breed is desirable if the lambs are short-legged, thick-bodied, and deep in the twist. Generally, black-faced or black-faced crossbreds are preferable to white-faced feeders unless too high priced. Remember, you are feeding lambs to make

SELECTING THE BREEDING STOCK

Fig. 2.16—A good bunch of feeder lambs. Notice that while of mixed breeding, they are alert, healthy, strong-bodied lambs which should make economical gains on good pasture.

money, so if wool breeds, for example, can be bought cheaply, they are desirable to feed out.

Consider the fleece—Lambs old enough to feed will have three to five pounds of wool. Therefore, an unshorn lamb is worth more than a shorn one. However, there is the cost of shearing, present price of wool, labor involved, etc., to contend with plus the fact that there will be less lamb (weight of wool) to sell later on. All of these factors must be taken into consideration in arriving at a proper price to pay.

Pick thrifty lambs—Fleeces should not be ragged and the animals should be alert and bright eyed. Thin, healthy lambs are better than fat lambs provided their thin condition is not due to parasites. Don't buy runts or under-sized lambs, as they rarely make proper gains.

Learn their origin—Of course, this is not always possible. However, if one knows what lambs have been fed, it can assist materially in selecting them. Lambs off irrigated pasture or those that have been grazed a long time are not the best bet. Whether or not internal parasites infest them can be determined easier if their past history is known, because some geographic areas are more infested than others.

Watch the shrink—It is best to buy lambs on a shrinkage basis. If impossible to hold off feed, the weight should be reduced three to four per cent to compensate for the fill.

4. Selecting Breeding Ewes

Many sheep breeders outside the range area make it a practice to buy all replacement ewes. This, of course, would not be true of breeders of purebred flocks, who would select the major numbers of their replacements within their own herd. Once started, many farm flocks can also be selected from their own stock. According to

Fig. 2.17—Large, rugged, uniform ewes such as these should be selected as the basis from which to build a herd.

SELECTING THE BREEDING STOCK

Virginia Polytechnic Institute, in most cases the cheapest way to bring about flock improvement is by using purebred males on the best ewes in the flock and keeping the best ewe lambs for replacements. Regardless of what plan is used, there are a number of approved practices to bear in mind when selecting ewes.

Select large ewes—The standards within the breed must be kept in mind, but generally speaking, within these standards large ewes are the most desirable provided they are not too coarse. Ewes must produce lambs if they are to yield a profit. Therefore, it is important to guard against barren ewes. Barren ewes usually are in much better condition than the general run of the flock from which they came.

Select desirable breed characteristics—This would certainly be true in purebred flocks, as color markings, grade of fleece, and conformation would be extremely important standards. However, even with range ewes or crossbreds such items as open faces, clean legs, grade of fleece, proper size, and conformation are important items to consider. It is most desirable that a group of ewes be selected on uniform standards and when once selected and segregated, they look even and alike in appearance. Straight legs, strong backs, deep bodies, and full twists, together with an alert, healthy general appearance, are desirable regardless of the breed type under consideration.

Buy young ewes—Farmers' Bulletin 840 says yearling or two-year-old ewes are preferable to older stock. Young ewes are still gaining in weight somewhat and much more apt to be healthy and produce good lambs and desirable fleeces. Two-year-old ewes that have lambed once, and have been sorted for soundness are the most desirable kind to get, but not always the cheapest. Watch for "best buys."

Check their mouths—Until a sheep is four years old

Fig. 2.18—This old ewe, purchased as a cull, is lame and has no teeth, yet she has produced four lambs in the last two years because of good care, and is about ready to lamb again.

its age can usually be told within a few months. **Kansas Bulletin 316** says in buying breeding ewes sheepmen should follow the rule that a ewe is just as old as her mouth. Frequently it is not the age in years that counts, but the age in condition of her mouth. For example, nine-year-old, solid mouth ewes are younger from the service standpoint than six-year-old, broken mouth ewes. The type of vegetation, gravelly soil, etc., are instrumental in causing teeth to wear more rapidly in some locations than in others. Any mouth unsoundness such as overshot or undershot jaws is reason for rejecting them. Occasionally old ewes can be obtained cheaply and made to yield one more lamb crop under good farm conditions and with extra care.

Examine teats and udders—Watch out for ewes that may be useless as breeders because of teats clipped off

SELECTING THE BREEDING STOCK

during shearing. Examine the udders to see that they are free from lumps that might prevent them from being milkers. Udders should be soft, spongy, free from lumps, warm to the touch, and both halves equal in size. Extra large teats and ruptures are undesirable.

5. Selecting Breeding Rams

Needless to say, too much stress cannot be put on the qualifications of a ram. Except in rare cases, it is a desirable practice to select purebreds. On the other hand, it would be a mistake to advise every sheep raiser to simply go out and obtain the most expensive ram he could find. Nevertheless, it is wise to buy the best you

Fig. 2.19—A desirable type Corriedale ram. Note the open face and rugged appearance. Corriedales have medium wool, high quality fleeces, and rugged bodies. The above animal was Champion ram at the 1969 Cow Palace. Rams should be selected according to your herd standard, and only those rams which show promise of improving the herd level used. (Courtesy, Dante Calvis, Bodega, California)

can afford. There are a number of approved practices that most good sheepmen will follow in selecting rams.

Obtain rams according to the quality of ewes—Always keep in mind the poorer points of your ewe flock and buy rams that may be able to correct these faults. For example, small, fine-boned ewes should be bred to rams with some size and ruggedness so that the lambs will have sufficient frame to grow out and finish for market. If your ewes are rather a poor lot, you don't need as good a buck as if they were very superior individuals.

Compare rams—It is easier to do a good job of selection if one has an opportunity to compare each ram against many others. Large, well established auctions are good places to do this as well as reputable purebred establishments which have a number of rams to select from. Good rams will "stand out" and their lambs will do likewise by going to market early at heavier weights.

Seek advice of others—County agents, experienced sheepmen, or other similar individuals can assist materially in helping the inexperienced obtain good rams. Many parts of the country have organized yearly ram sales to which all good purebred breeders consign rams. This is an excellent source of breeding rams, as sifting committees of capable people go over the entire offerings and eliminate any animal not up to standards. Not only does one have the advantage here of comparing many animals, but also of seeing large numbers of good animals in one location, thus doing away with a great deal of travel.

Buy young rams—Yearlings are preferable when it is necessary to buy a ram. According to **Kansas Bulletin 316,** yearlings are usually more vigorous and can be used for a longer time. Occasionally, tried rams can be obtained at the right price and they are desirable. Ram lambs may sometimes be used to a very limited

SELECTING THE BREEDING STOCK

extent if large and well grown, but most sheepmen do not favor this practice.

Select healthy, well-formed animals—It goes without saying the rams should be free from internal and external parasites in so far as one can determine. They should be disease free, vigorous, and guaranteed breeders. Reputable breeders will guarantee their stock in this respect, so buy from those who do. Faults in conformation, such as crooked legs or parrot jaws, should be guarded against as there is always the possibility of these deficiencies being transmitted to the offspring.

Buy early—Don't wait until the last minute to find a ram. Some breeders will keep rams until the buyer has a place for them. On the other hand, it is a desirable practice to get new rams accustomed to new environmental and feed conditions so they will be ready for the breeding season. According to **California Extension Circular 49,** a good rule is to purchase rams a month or two before turning them in with the ewes.

6. Buying at the Right Time

The time of year and the price paid for breeding stock are prime factors in determining if a sheep enterprise is to be successful. There are a number of approved practices one should follow in getting established in the business.

Start at right time—Late summer or early fall is the most favorable time to make a start in sheep raising. **Farmers' Bulletin** 840 says ewes can be procured more readily at this time, and when purchased can be kept on meadows, grain stubble fields, or late sown forage crops to get them in good condition for breeding. It is seldom possible to buy any considerable number of bred ewes reasonably. In addition, the experience gained through fall and winter with the ewes, as well as getting them used to new surroundings, will stand the beginner in good stead at lambing time.

Buy grade ewes—The raising of purebreds and selling of breeding rams can best be undertaken only after considerable experience is gained. Therefore, the inexperienced sheep raiser should begin with the best grade ewes obtainable and a purebred ram.

Pay market price—This may sound like strange advice, but good ewes are seldom a bargain. On the other hand, purebred ewes may often be too high priced for their quality. Ewes from the western ranges can often be obtained directly from a stockyard market. According to **Illinois Circular 657**, a good, vigorous ram that may be mated to 40 ewes is easily worth five times or more as much as the average ewe in the flock. There-

Fig. 2.20—Scene at a California ram sale. Events such as this bring together many good animals where sifting committees eliminate most unworthy individuals. As a general rule, a sound market price is paid for each individual animal.

SELECTING THE BREEDING STOCK

fore, you would expect to pay much more for a ram, but not more than going market price for similar type rams.

Learn the market—Radio, newspapers, magazines, and livestock journals all carry current market reports and sales. Those in the sheep industry should keep up on present prices and the market trend. Commission firms at central markets can inform prospective buyers of market conditions and the latter should not be overlooked. Reputable order buyers can often obtain good ewes at reasonable prices.

Watch the show dates—Those who wish to enter fat lambs for show must purchase or raise lambs that will be at acceptable weights and ages for the particular show they wish to enter. This, of course, will vary somewhat for each state or area. In Oklahoma, for example, there are two show seasons—the spring show in March and the state fair in September. Many states follow a calendar similar to this. According to **Oklahoma Circular 468,** members should select lambs that will be less than one year of age at the time shown. Lambs born in January and February are best to feed for September shows, whereas late lambs in May or June are best for feeding for the following March shows. In California, for example, most lambs shown in the spring shows in April or May are born the previous November or December. Finished lambs are difficult to hold in show bloom, so it is important to get them at the right age so they will be at the proper weight by show time and will not have to be held back. It is a desirable practice to get a premium catalog for show dates and check back to find the proper age to select lambs.

7. Selecting for High Productive Capacity

For a flock to be profitable, each ewe must regularly

WHITEFACE DUAL-PURPOSE SHEEP	
Size and conformation, 70 per cent	**Per cent**
Weight for age and scale	40
General appearance and breed type	4
Shoulders, chest and spring of ribs	5
Back and loin	3
Rump and leg of mutton	10
Natural fleshing	3
Feet and legs	5
TOTAL	70
Fleece, 30 per cent	
Wool grade (breed considered) and uniformity (all parts of body)	12
Length of staple	12
Density	3
Character (crimp, color of secretions, freedom from hair)	3
TOTAL	30

BLACKFACE MUTTON SHEEP	
Size and conformation, 90 per cent	**Per cent**
Weight for age and scale	50
General appearance and breed type	7
Shoulders, chest and spring of ribs	7
Back and loin	4
Rump and leg of mutton	12
Natural fleshing	5
Feet and legs	5
TOTAL	90
Fleece, 10 per cent	
Wool grade (breed considered) and uniformity (all parts of body)	2
Length of staple	4
Freedom from black fiber	2
Belly wool	2
TOTAL	10

Fig. 2.21—Score cards used in grading sheep. Courtesy, California Agricultural Experimental Station Manual 40)

SELECTING THE BREEDING STOCK

produce good, market lambs and an abundance of high quality wool.

Cull carefully—Constant selection and careful culling of ewes are necessary to reach high standards. Any ewe that does not meet the standard of the flock should be culled. Ewes not producing a lamb should be culled. However, those which are difficult breeders and produce odd season lambs, have poor fleeces, are chronic bloaters, or poor feeders, etc., should be eliminated as well.

Train your eye—As one gains experience, he can easily select those desirable animals he wishes to keep. Until he does so, it is advisable to study the score card and try to apply it in learning to select sheep.

8. Deciding on Number to Stock

This decision is an important one, not only for the beginner but the experienced sheepman as well, because of the danger of overstocking. **Texas Agricultural Extension Service B-827** says:

"The number of sheep that a farm will carry on a year-round grazing plan depends on the size of the pastures, the amount of rainfall, the fertility of the soil and the amount of supplemental roughage available.

"Many farmers know the carrying capacity of their farms in terms of cattle. Ordinarily five to seven mature sheep will replace one cow, depending on the size or breed of the sheep. This does not mean in addition to the cattle.

"It is best to start with fewer sheep than the farm can carry and grow into the proper number as experience is gained. A few good quality ewes are more profitable than a large number of poor quality ewes. High production per ewe of both wool and lamb is essential to a successful operation."

CHAPTER III

BREEDING AND IMPROVING SHEEP

Raising the standards of the sheep industry through sound breeding and improvement practices is one of the most challenging aspects of the entire enterprise. This is true whether it is considered from the standpoint of the industry as a whole or from the standpoint of raising the level of an individual flock. While there are many excellent flocks throughout the country, the average standard is surprisingly low. Percentage lamb crop, e.g., should be well over 100%, yet the average is around 88%. By the use of approved practices, these figures and many other standards can be raised to a more profitable level.

Activities Which Involve Approved Practices

1. Selecting a method of breeding.
2. Maintaining a high percentage lamb crop.
3. Breeding at the right age.
4. Methods of mating.
5. Culling the flock.
6. Obtaining replacements.
7. Caring for the ram.
8. Progeny testing.
9. Summary of practices to improve flocks.

1. Selecting a Method of Breeding

A great many avenues are open to sheep breeders in regard to different methods of breeding. New breeders in particular should be careful to select a sound breeding program that fits their own climate and farming conditions.

Don't develop a new breed—It is true that periodically new, improved breeds of sheep have appeared. However, in each case it has been the result of long, hard, painstaking effort by very experienced sheepmen or professionally trained animal husbandmen. Even in these cases, many so-called new breeds fail to come up to the expectations of their creators. Most sheep raisers recommend staying with a recognized breed. There is a good breed available for almost every known farm condition. (See the chapter on breeds.)

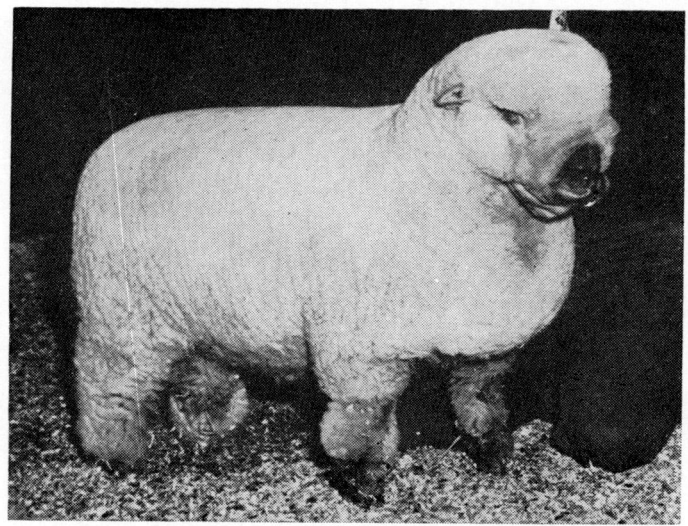

Fig. 3.1—A fine type of Shropshire ram capable of improving the standard in many herds. Purebred rams should be used even on native ewes.

Use purebred rams—This is obvious for purebred breeders. However, most range sheepmen and others with good flocks find it a desirable practice to use purebred rams even on native ewes.

Stay with one breed—There are, however, specialized conditions which would make this advice unprofitable. These conditions will be discussed later. Nevertheless, most farmers and range producers find it advisable to stay with one breed. Improvement comes by selecting and culling within the breed. Replacement stock and rams are easier to obtain than if more than one breed is kept.

Crossbreed under specialized conditions—Producing spring lambs, high quality feeder lambs, and hothouse lambs is more profitable when carried out under a crossbreeding system. The particular breed of ewe used determines when the lamb can be produced; in most cases, early, and in addition, generally yielding a heavier clip. Mutton-type bucks give larger, heavier, quicker maturing lambs coupled with a certain amount of hybrid vigor. Such feeders, resulting from crossbreeding, are in top demand for feedlot fattening or to finish on irrigated pastures.

There are four systems, according to **Animal Science Mimeo, Series 116,** of the Ohio Agricultural Experiment Station, which may be employed in crossbreeding as clearly defined programs:

1. Direct crossing.
2. Systematic progressive crossing.
3. Rotation crossing.
4. Crisscrossing.

Direct crossing refers to the crossing of two types or breeds. A good example is the crossbreeding of mutton-type rams to wool-type ewes or, let us say, Hampshire ram $\times$ Merino ewe. The ewes for this system usually are provided by maintaining half the flock as "source" or pure stock, or the ewes are available for purchase.

Systematic progressive crossing refers to the use of three or more breeds in an established order. This system utilizes crossbred females; first the F_1, next the three-breed-cross ewe, then continuing for as many different crosses as are planned. For example, Columbia rams may be crossed with Rambouillet ewes, the F_1 females may be retained and bred to Hampshire rams; the three-breed-cross ewe in turn is bred to a ram of a fourth breed, and so on until the program is terminated. The designation for such a four-breed-cross might read: Suffolk $\times$ Hampshire $\times$ (Columbia $\times$ Rambouillet).

Rotation crossing refers to the successive use of three or more breeds or types of ram, starting with the base ewes, and utilizing in each turn the crossbred ewe, ultimately backcrossing to the ram of the initial breed and following through again in the same order as in the initial part of the program. Each cross furnishes the ewes for the next cross.

Crisscrossing refers to back and forth crossbreeding using only two breeds. This system furnishes the ewes at each stage for the next step and, theoretically, could be continuous. However, essentially the same effect can be obtained through outbreeding within a single breed; that is, selecting a ram each time that is totally unrelated to the females in use.

Purebred breeders study specialized methods—More intensified methods must be used by those who have excellent type individuals and wish to maintain a supreme place in sheep breeding circles. Some of the methods used are:

Outcrossing — Mating relatively unrelated animals within the same breed. Such matings have no common ancestors within the first four to six generations. Excellent sheep of the same breed from different parts of the country are obtained in order to maintain the program.

Inbreeding—Mating of closely related animals. Inbreeding is divided into two parts—closebreeding and linebreeding.

Closebreeding—Mating of sire to daughter, son to dam, full brother to sister.

Linebreeding—A more distant form of inbreeding—half brother and sister, cousins, etc.

Linebreeding is the most common method of inbreeding. While these intensified methods are necessary to isolate desirable characteristics and then concentrate them into one strain, it is not without its dangers. Therefore, rigid culling must be employed under such conditions. It is a desirable practice that only the experienced and those possessing real know-how in the science of genetics utilize these methods.

2. Maintaining a High Percentage Lamb Crop

The most important single factor in making a profit in the sheep business is the percentage lamb crop. It is doubly significant because it can be influenced to a considerable extent by the practices the owner uses. **Cornell Bulletin** 828 states that large flocks should have at least a 100% lamb crop and it should be even higher for smaller flocks. There is some breed difference, but many sheepmen regularly have 130% to 160% lamb crop, or even higher.

Produce one lamb crop a year—It is possible for some ewes to produce two lamb crops in one year. However, most experienced shepherds know that one good crop a year is best from many standpoints, such as the

Fig. 3.2—This Suffolk ewe gave birth to four lambs and is raising them, a rare occurrence. It is important for ewes to raise twins in order to maintain a high percentage lamb crop. (Courtesy, David Starr, Altures, California)

ewes' health, management factors, and the availability of milk producing feed.

Flush ewes prior to breeding—About 15 to 20 more lambs per 100 ewes may be expected by flushing, according to **Bulletin 68** of the Ohio State University. Flushing is accomplished by forcing the ewe to lay on flesh from 10 to 20 days before being turned in with the ram. Under such conditions the reproductive organs generally produce more eggs. The key to flushing is not that the ewes be fat, but that they be in good health and gaining in weight. This can be accomplished by adding a concentrate to the ration or putting them on a better pasture than they were before.

Condition the rams—A heavy fleece and high condition frequently result in lower fertility or in sterility. Show or sale rams should be let down gradually in flesh. Rams should be brought to new surroundings early, in time to get accustomed to them and to develop a vigorous condition. If rams are in poor flesh, they should be fed a good ration of legume hay or allowed to run on grass and fed one-half to three-fourths pound daily of a concentrate. Oats alone, or oats and bran two to one, make a good concentrate mixture.

Shear rams—Shear them all over if they are to be used early. At least shear legs, neck, brisket, belly, and scrotum. Rams will be more active and it is easier for them to cover a ewe if they are shorn.

Use correct number ewes per ram—Under rough range conditions, three rams per 100 ewes is a desirable ratio. Yearling rams can handle from 25 to 40 ewes. However, mature, strong rams may handle up to 75 ewes although 35 to 50 is most desirable under farm conditions.

3. Breeding at the Right Age

Breed to lamb at two years—Ewes will come in heat every 16 or 17 days during the breeding season. They

Fig. 3.3—These rams have just been shorn. As they are to be used shortly for breeding, they will be in active condition.

stay in heat one to three days. Lambs are born about 147 days (21 weeks) later, although the gestation period may vary from 142 to 152 days. It is important that ewes be well grown, as undersized ewes, or ewes lambing the same year they are born, frequently disown lambs or have difficulty lambing.

Breeding Hampshire ewes for the first time as lambs resulted in a material increase in total lamb production with only a slight decrease in wool production. Early

GESTATION TABLE FOR SHEEP

Date of Service	Date Animal Due to Give Birth	Date of Service	Date Animal Due to Give Birth
	Ewe		Ewe
Jan. 1	May 31	July 10	Dec. 7
Jan. 11	June 10	July 20	Dec. 17
Jan. 21	June 20	July 30	Dec. 27
Jan. 31	June 30	Aug. 9	Jan. 6
Feb. 10	July 10	Aug. 19	Jan. 16
Feb. 20	July 20	Aug. 29	Jan. 26
Mar. 2	July 30	Sept. 8	Feb. 5
Mar. 21	Aug. 9	Sept. 18	Feb. 15
Mar. 22	Aug. 19	Sept. 28	Feb. 25
Apr. 1	Aug. 29	Oct. 8	Mar. 7
Apr. 11	Sept. 8	Oct. 18	Mar. 17
Apr. 21	Sept. 18	Oct. 28	Mar. 27
May 1	Sept. 28	Nov. 7	Apr. 6
May 11	Oct. 8	Nov. 17	Apr. 16
May 21	Oct. 18	Nov. 27	Apr. 26
May 31	Oct. 28	Dec. 7	May 6
June 10	Nov. 7	Dec. 17	May 16
June 20	Nov. 17	Dec. 27	May 26
June 30	Nov. 27		

Ewes are in heat from 1-2 days. If not impregnated, heat will recur in 17 to 28 days.

breeding appears to be economical and profitable under conditions where ewe lambs can develop adequately and where Hampshires or similar breeds can be given extra feed and care.

4. Methods of Mating

Proper methods of mating mean a uniform lamb crop which not only results in higher income, but cuts down on expense and labor at lambing time as well.

Maintain a short breeding season—Rams should remain with the ewes only six to eight weeks. Therefore, it is important to cull so that proper type ewes will be maintained and thus insure a short breeding season. Raising one's own replacements tends to insure a shorter breeding season than ewes obtained from widely separated areas. It may take several years to develop a short lambing period, but it is a practice worth the effort.

Rotate the rams—Common practice is to allow the ram to run with the ewes during the breeding season. However, if the number of ewes is large, or a young ram or old ram is used, better results are obtained if he is penned in the day and allowed to run with the flock only at night. He will soon learn to leave the flock for a pan of grain. When a large band is run together, as on western ranges, a good practice is to divide the rams into several groups and rotate the bucks about once a week so they can rest up in the meantime.

Mark the ram—A mixture of yellow ochre and linseed oil should be smeared on his brisket. Although it may have to be put on every day or two, it will enable the owner to tell which ewes are bred and whether or not the ram is mounting them. According to **Ohio Bulletin 68**, on the eighteenth day after marking the buck, it is advisable to change to another color, e.g., venetian red and linseed oil, to see whether or not the ewes are returning in heat. If such is the case, it may be advantageous to change rams, although failure to settle the ewes is not always the ram's fault.

Fig. 3.4—A good time to cull ewes is at shearing time as the amount of wool can be measured and body conformation easily seen.

5. Culling the Flock

Mutton and wool yields can be increased over a period of years by setting up a rigid system of culling. Purebred breeders may adopt a more severe method of culling than average range or farm flock breeders. However, each would use the same general standards and approved practices.

Cull at shearing time—Although culling is really a year-round job, special attention should be given at shearing. This is an especially desirable practice for range sheep men or large farm flocks for two reasons. First, by having a set time, no sheep escape a close scrutiny to see if they are up to standards. Second, the wool has just been removed and thus can be graded and weighed. In addition, body conformation can be seen more easily and evaluated to decide whether or not each animal is up to the standard of the flock.

Cull on production—Each ewe should produce a strong, healthy lamb plus the proper weight and grade of wool for her particular breed. Udders should be in good health and free from lumps. Mouths should be inspected, particularly on range sheep, in order to determine if they can go through another season and still remain in good condition. According to **University of Illinois Circular 657**, a simple record showing the number of the ewe, weight of fleece, number of lambs dropped and raised, and weight of lambs at weaning will suffice to give enough information so that ewes can be culled on production.

Identify ewes—Farm flocks and purebred herds can do a better and more systematic job of culling year after year if all ewes are easily identified. Every ewe should have a metal tag or ear label bearing her flock number. The same is true of rams and ram lambs in purebred flocks.

Fig. 3.5—Thin bodied, crooked legged ewes like this one should be culled from the breeding herd.

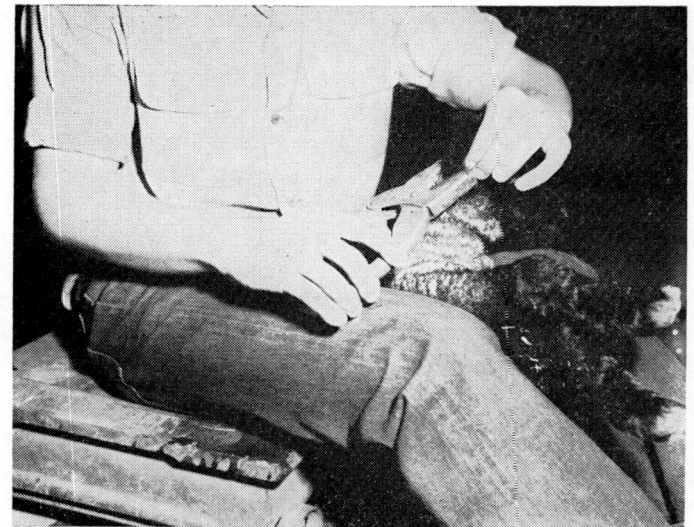

Fig. 3.6—Placing a metal tag in the ear of a lamb. Proper identification is essential to systematic culling.

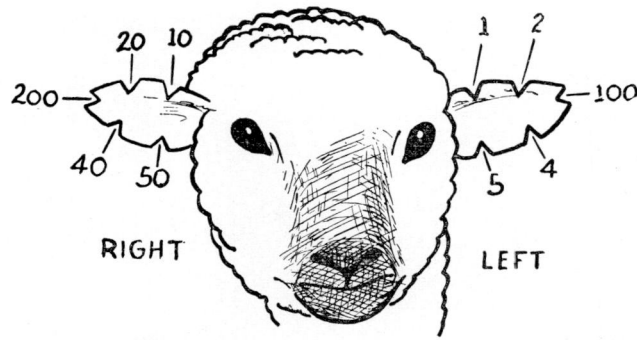

Fig. 3.7—One system of numbering sheep by use of notches cut in the ears. Combinations of the above notches will give any desired number up to 400.

Cull according to flock standards—A cull in one flock may not be a cull in another flock of lower standards. For example, a cull in a high-class, purebred herd may be a desirable animal in the average farm providing, of course, general health and the ability to produce a lamb is evident. Therefore, each breeder must determine the standard for his flock and cull accordingly.

6. Obtaining Replacements

To maintain the number and the right average age of the band, the sheepman must replace ewes that are lost or culled each year with an equal number of young ewes. It is advisable to improve the average quality of the flock each year by disposing of the poorer animals and replacing them with better quality young stock.

Keep proper number of ewe lambs—In a flock of 100 ewes, 18 or 20 of the best ewe lambs should be kept for replacements. According to **Ohio Extension Service Bulletin 68,** this will allow for replacement by death, ewes removed because of age, and some culling of the younger breeding stock. Mutton ewes start on the decline at six years of age and wool breeds at seven years. Range sheepmen prefer to cull before this age because of the rigorous life range sheep encounter.

Buy western range ewes for crossbred production—It is generally considered advantageous to buy replacements from the western ranges unless one has a purebred flock or is raising sheep of one breed. Those breeders producing a special market lamb by crossing blackfaced bucks with white-faced ewes can best replace their ewes by obtaining them from western range flocks where desirable white-faced wool breeds predominate. Reliable order buyers or farmer cooperatives can assist in making wise purchases.

Raise your own replacements—This practice should not be overlooked for farm flocks where one breed predominates. Improvement is obtained by purchasing excellent type rams from good purebred flocks so that lambs tend to be better quality animals than their

Fig. 3.8—A desirable group of replacement ewes.

mothers. Purebred breeders use this practice almost exclusively.

A more scientific breeding program can be utilized by selecting replacements from your own herd as records, longevity of parents, etc., can be used by the sheepmen in deciding most likely matings. In addition, selection can be based on other factors, such as gain in weight, rather than selection by body conformation alone as often must be the case when replacements are outside a sheepman's own herd.

7. Caring for the Ram

It is a shortsighted policy to use inferior or off-type rams. Over a period of time the ram is at least half the flock. Rams should be selected that have their strong points where the ewes are most lacking. If this is true, then it is especially worthwhile to use approved practices in caring for the ram.

Keep ram in vigorous condition—This is particularly important during the breeding season. Grain feeding may be necessary during the breeding season in order to maintain his vigor. Rams should not be cooped up as they tend to become too fat and sluggish without adequate exercise. A good procedure is to run the ram with other rams or several wethers in a pasture or large lot so they can romp together and maintain good muscle condition.

Shear rams—Sale and show rams should be shorn in preparation for the breeding season. Shearing the ram will cause him to be more active, decrease the likelihood of temporary sterility, and make it easier for him to cover a ewe. At least the legs, neck, brisket, belly, and scrotum should be shorn out. Oftentimes this practice will be unnecessary as the breeding season will follow normal flock shearing operations closely enough.

Maintain medium flesh—This condition will keep the ram most active during the breeding season. A medium-sized ram will keep in flesh nicely during the breeding season on a pound a day of some grain mix in addition to good hay or pasture. Three parts oats and one part wheat bran by weight is a desirable mix. Grazing on good legume pasture is an excellent method of keeping the ram in good flesh, although he probably will need grain during the breeding season. Guard against too heavy feeding, as evidence indicates overweight rams may be sterile.

8. Progeny Testing

Progeny or performance testing refers to the use of information about offspring of animals to determine how animals shall be mated in order to more accurately breed for improved livestock. It must be borne in mind that this method is only another tool in a progressive breeding livestock program and does not supplement individuality and pedigree selection. It is, however, a more scientific approach to improved breeding, particularly from the standpoint of establishing desirable qual-

Fig. 3.9—Good breeding and feeding pay yield results as indicated by this three-year-old ewe and her second lamb. This Merino ewe weighing 110 pounds shorn, grew 16.25 pounds of light-shrinking wool with a 4.5 inch staple in 12 months. Her four-month-old lamb weighed 57 pounds.

ities such as rate of gain and efficiency in use of feed or other qualities not so readily observed by the eye.

Keep accurate records—The key to progeny testing is, of course, accurate records. Regular testing and recording of such information as fleece weight per measured period, weaning weight, number of lambs raised, and the like, must be correctly recorded and standardized so that animals can be properly compared. For example, if weights of lambs are to be compared, all weighings must be made for the same length of time so that one lamb is not older than another even if only by a few days. In addition, such things as ewes being on similar feed and grazing conditions must also be taken into account. If such were not the case, obviously comparisons could not be justified.

BREEDING AND IMPROVING SHEEP 85

Use records fairly—Oftentimes there is a tendency to disbelieve records and let personal preference and prejudice influence which matings are to be made and how animals are to be culled. However, if performance records are to be of value, they must be used fairly, and matings and culling done according to what the records show in spite of the fact that a favorite ram must occasionally be culled.

Fig. 3.10—A convenient way to weigh lambs. First adjust scale for weight of bucket.

LAMB PRODUCTION RECORD

Tag _____

Year											
Date ram turned in											
Date out											
Ram used											
Ram grade											
Date lambed											
Lamb No.											
Sex											
Birth weight											
Raised on (ewe No.)											
Date castrated											
120-day weight											
120-day grade											
Sale price											
Disposition of lamb											

WOOL RECORD

Year								
Date shorn								
Weight								
Days growth								
Staple length cm								
Grade (fineness)								
Remarks (quality, etc.)								

HERD BOOK RECORD

Date lambed................Type birth............Reared as............Reg. No. ....Breed............Notch......Tag........

Dam No.............Age of dam.........Dam's grade............Sire No............Sire's grade............Sex........

Birth wt............ Weaning wt............

Remarks.

This is a page from a recommended herd record book. The amount of information to keep on each animal will depend upon type of breeding program and analysis desired.

Fig. 3.11—An open-faced ewe lamb. Within the limits of the breed, a reasonable standard is to breed for open-faced animals.

Decide on reasonable standards and ideals—Breeders must first have a desirable objective and ideal in mind. Without such goals, the program will not progress in any particular direction. These goals should be logical and, while high, should also be within reach and tangible. If a breeder's ideals are too high or unattainable, it is possible to become frustrated and develop a sense of futility regarding a breeding program. Goals, of course, may vary, as breeders starting with excellent type sheep will probably select goals higher than one starting with only an average flock. Furthermore, standards may change either because of change in mar-

BREEDING AND IMPROVING SHEEP

ket demands, or as present standards are reached, sheepmen will set higher ideals.

9. Summary of Practices to Improve Flocks

Individual producers will find **Chart MP-244**, developed by Texas Agricultural Extension Service for county survey, extremely helpful in evaluating their own breeding and improvement programs.

APPROVED SHEEP PRODUCTION PRACTICES

Practice	Recommendation	County Situation		
		Good	Fair	Needs Improvement
1. Select a breed that has proved profitable for the county or area.	Consider adaptability and type of production.			
2. Follow a sheep selection program.				
a. Age.	Cull according to the productivity of the sheep.			
b. Size.	The size should be representative of the sheep of the area.			
c. Conformation.	Conformation is necessary to produce a desirable market lamb.			
d. Open face.	Open-faced sheep produce more pounds of lamb.			
e. Length of wool.	Fine wool sheep should produce ¼ inch of wool per month or 3 inches per year.			
f. Fineness of wool.	Fineness of wool should be representative of breed.			

(Continued)

APPROVED SHEEP PRODUCTION PRACTICES (Continued)

Practice	Recommendation	County Situation		
		Good	Fair	Needs Improvement
g. Uniformity of fineness.	Discriminate against hairy britch or lack of uniformity.			
h. Completeness of covering.	The body should be well covered with wool except for the face.			
3. Select rams from high-producing flocks.	Use production tested rams when possible.			
4. Practice seasonal breeding.	Follow practices that best fit the county.			
5. Control external parasites.	Spray or dip with recommended insecticides.			
6. Control internal parasites.	Drench as needed; use phenothiazine salt.			
7. Control diseases.				
a. Soremouth.	Vaccinate baby lambs.			
b. Bluetongue.	Vaccinate at 4 to 5 months and as needed.			
c. Enterotoxemia.	Vaccinate as needed.			
8. Control predators.	Employ a government trapper; enact and enforce a county dog law.			

(Continued)

BREEDING AND IMPROVING SHEEP

APPROVED SHEEP PRODUCTION PRACTICES (Continued)

Practice	Recommendation	County Situation		
		Good	Fair	Needs Improvement
9. Castrate and dock.	Perform operations from 1 to 3 weeks of age.			
10. Provide protection from weather.	Provide a shed or natural shelter.			
11. Feed mineral supplements.	Feed minerals to meet county needs.			
12. Provide the proper number of rams to ewes.	Provide a minimum of 3 rams per 100 ewes. Large, rough or brushy pastures may need more rams.			
13. Provide adequate watering facilities.	Sheep should not have to travel over ½ mile to water.			
14. Distribute salt properly.	Salt away from water for better utilization of pastures.			
15. Stock pastures properly.	Follow the stocking rate recommended for the county.			
16. Provide supplemental feed.	Use home-grown feeds when possible; buy supplemental feeds when they are plentiful.			
17. Increase the per cent of lamb crop.	Practice better management at lambing time.			

CHAPTER IV

HANDLING SHEEP AND LAMBS

While sheep are not strong animals capable of harming humans and therefore necessitating specialized equipment, they require a great many specialized operations for their care that other animals do not. Shearing and docking, for example, are specialized skills necessary for their successful handling that is unnecessary in handling any other type of livestock. In fact, their very helplessness and lack of strength make it particularly important that sheepmen learn the fundamental skills in handling animals so that large numbers may be taken care of easily, efficiently, and without injury to the sheep. Damage to the wool or the skin or suffocation can mean a serious financial loss unless sheep are handled properly.

Activities Which Involve Approved Practices

1. Shearing sheep.
2. Grading wool.
3. Examining the fleece.
4. Utilizing the gregariousness of sheep.
5. Winter handling.
6. Setting up sheep.
7. Marking sheep.

8. Tagging sheep.
9. Trimming feet.
10. Painting rams and branding sheep.
11. Telling age of sheep.
12. Tattooing sheep.
13. Training show sheep.
14. Diagnosing pregnancy.

1. Shearing Sheep

Quality plays the most important part in determining the value per pound of wool. However, a number of clean, neatly tied fleeces present a pleasing picture to the flock owner and wool buyer. Not only is appearance important, but condition of wool, the presence or absence of foreign material, and kind and amount of wool twine used are also significant in producing a neat appearing clip.

Truck small bands to commercial shearers—Under this system a fixed price per head is paid, but it has the advantage, generally, of an expert person shearing the sheep. Good shearers are easier on the sheep, make fewer second cuts, and keep the fleece in one piece. Trucking sheep generally means cleaner sheep on arrival and a lower price than if the shearer must come to the farm where his equipment is not set up.

Learn to shear your own sheep—This practice may seem controversial to the preceding advice. However, the small sheepman is always plagued with the lack of experienced shearers available when he needs them, with the result that a great deal of expensive running around takes place in order to find qualified shearers. If the owner of a small flock would take the trouble and time necessary to learn to shear sheep, not only would he solve his own problem, but often the problem of his close neighbors. In addition, the income from shearing several neighbors' flocks will justify the expense of owning power equipment. The experience one gains in

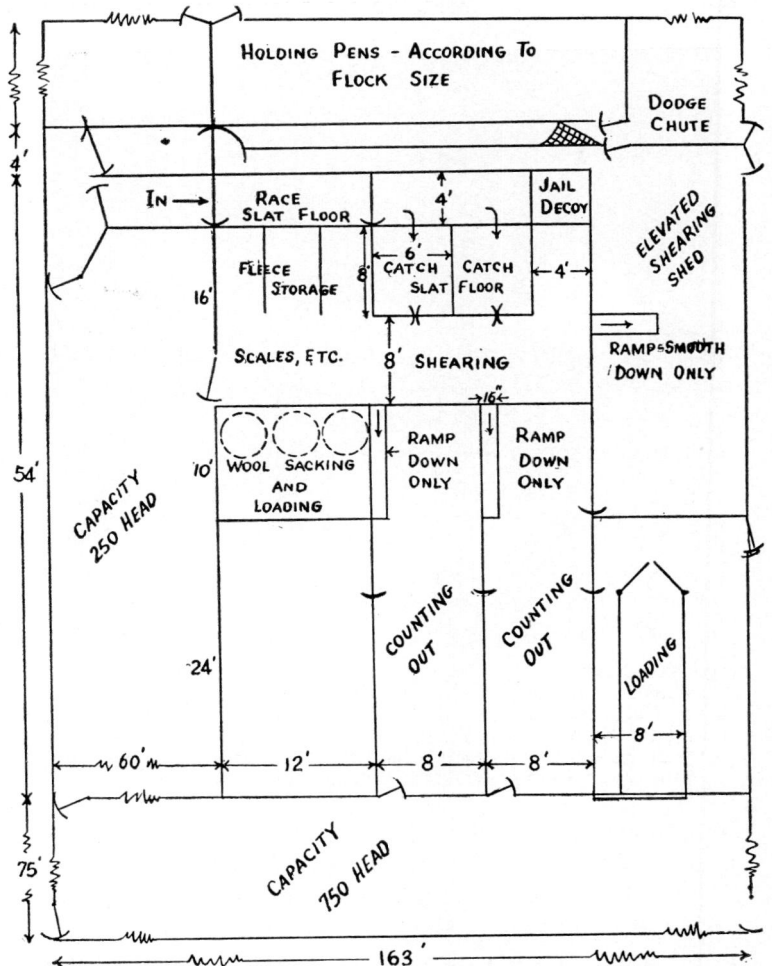

Fig. 4.1—This shearing pen layout is recommended for large operators of 800 to 1200 head. The shearing shed is elevated so droppings, urine, and dirt will fall through the slatted floor of the race and catching pens, thereby keeping sheep clean for shearing. In addition, the raised shed permits sheep to stay underneath at night so fog and dew on the wool will not be a problem. Ramps are smooth so sheep must go down and cannot back up. Central arrangement of sheds gets sheep used to going in and out so herding is no problem. (Courtesy, Don R. Richardson Ranch, Stewarts Point, California)

Fig. 4.2—Farmers are learning to shear sheep at a shearing school conducted by the Sunbeam Corp. and the University of California. In two days the average farmer can master all of the necessary fundamentals of shearing. An excellent chart on shearing sheep is available from the Sunbeam Corp., 5400 West Roosevelt Road, Chicago, Illinois 60650.

doing many sheep will make him a more expert shearer as well. One disadvantage, of course, is that shearing is hard labor even though it comes for just a brief period every season. Shearing is a tricky skill to acquire; however, agricultural colleges and county agents frequently hold shearing demonstrations and practice sessions where farmers can learn to shear sheep. The Sunbeam Corporation will supply excellent charts and diagrams to guide beginners in learning to shear. Almost any person able to handle sheep can learn to do a creditable job of shearing in one season.

Plan convenient sheds—The better types of shearing

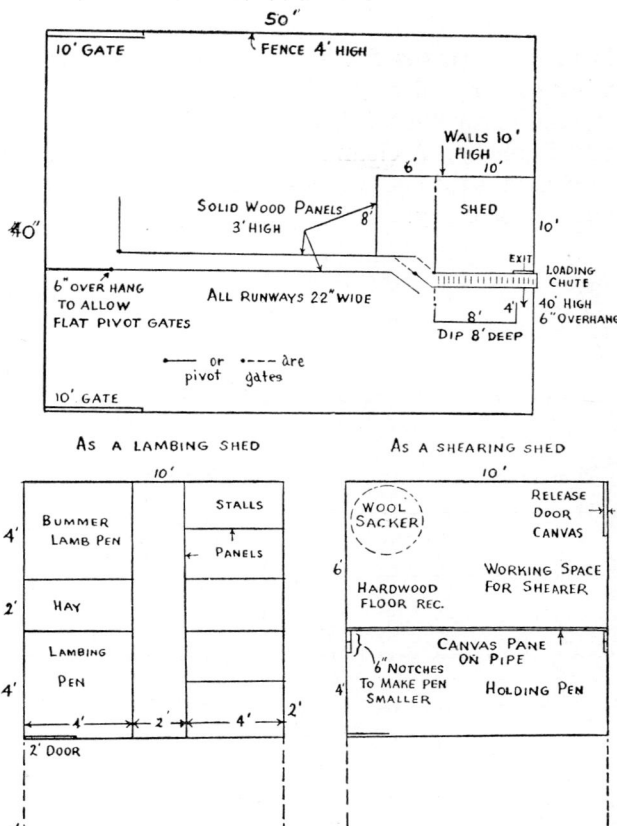

Fig. 4.3—A combination lambing and shearing shed recommended for small producers up to 300 head. The two enlarged drawings in the lower part of the diagram illustrate how the shed is converted into either a lambing or shearing shed. The panels in the shed are all removable either to fit a different function as shearing, lambing, or storage or to fit the size of the sheep. Note the size of the pens, 2 by 4 feet; these would be too small for a large breed like the Hampshire or Suffolk if they were simply to be confined to be observed or to lamb normally. On the other hand if lambs are being "grafted" onto a new mother, with the use of a sack harness and 2 x 4 wooden pole, then the pens must be small enough so she cannot turn around. Once the ewe accepts the new lamb, generally in a day or two she can be let out or the panels removed or shifted to enlarge the pen. If a ewe is restrained long enough for the lamb to nurse and the milk pass through its digestive tract, the smell from the resulting feces will convince the mother the lamb is her own. (Courtesy, W. L. Juergenson Ranch, Auburn, California)

sheds provide means for sweating the sheep to make shearing easier and convenient pens for holding sheep before and after shearing without mixing sheared and unsheared sheep. A clean space for shearing the sheep and suitable equipment for sacking and storing the wool are a necessity.

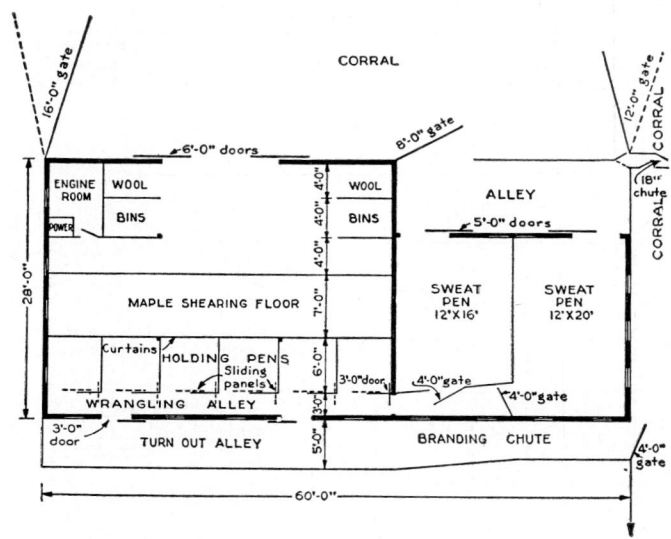

Fig. 4.4—Plan of a shearing and sweating shed recommended by the United States Department of Agriculture for very large sheep ranches.

Use hardwood floor—A good hardwood floor such as a maple floor is best where shearing is actually to take place as it prevents dirt and splinters from catching in the wool. When such a platform is not available, a canvas spread on the bedded sheep pen will serve satisfactorily.

Shear after warm weather begins—It is best to shear after the weather is warm enough so newly sheared animals will not suffer from exposure. According to **Farmers' Bulletin 1710,** wet snowstorms and severe

Fig. 4.5—An expert shearer properly shearing a ewe. Note the smooth hardwood floor so dirt and splinters will not catch on the wool.

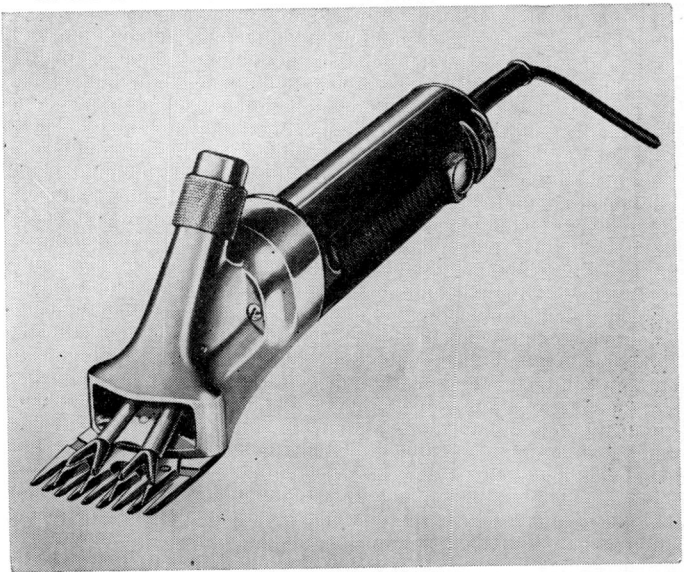

Fig. 4.6a.—A small electric type of machine shears suitable for tagging and shearing small farm flocks.

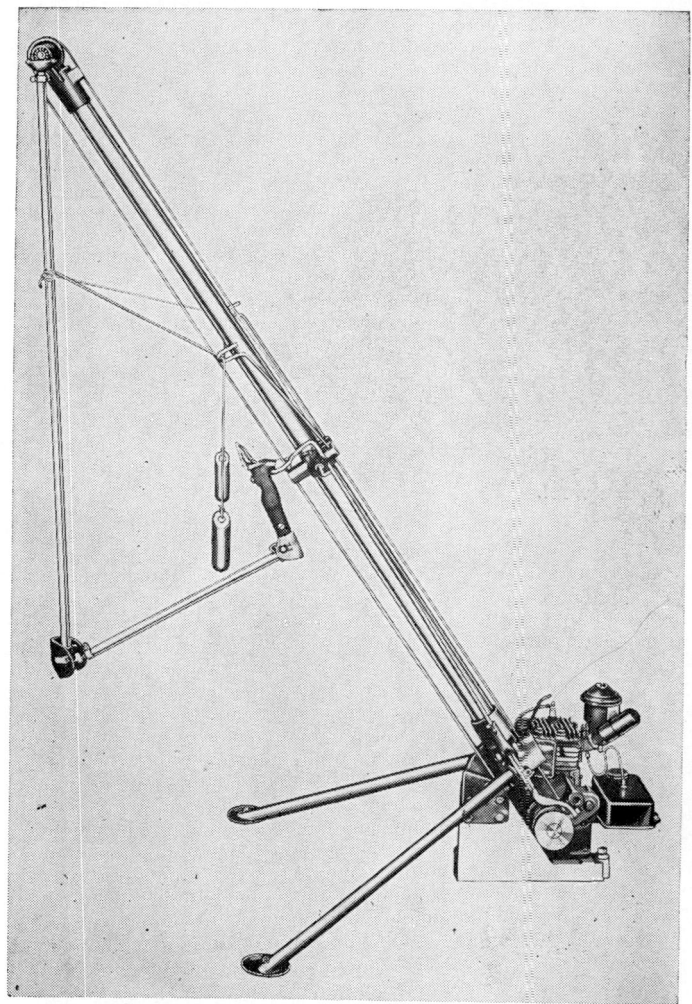

Fig. 4.6b—A larger type of machine shears used by commercial shearers. (Courtesy, Sunbeam Corp.)

HANDLING SHEEP AND LAMBS

cold are sure to cause heavy losses among newly sheared sheep. Some sheepmen shear before lambing, but the majority favor waiting until after lambing, as there is danger from mechanical abortion. In the western range country, most animals are sheared during May and June. However, in the warmer valleys of the Pacific coast and in the south, shearing may begin in March. It is advisable to shear twice a year if weeds, such as foxtail, are a major problem, even though it reduces the value of the clip.

Machine shear—Hand shearing was very popular in the past. Even though many sheep are still sheared in this fashion, most sheep are now sheared by machine. The advantage of machine shearing is that the wool is taken off more closely and uniformly. There are fewer second cuts, and the entire operation is quicker, thereby getting more sheep done per day. In addition,

Fig. 4.7—Remove the fleece in one piece. Notice how the shearer's legs assist in holding the ewe.

it is easier for people to learn to do and not as hard labor. Of course, the number of sheep per operator per day is greatly increased.

Remove fleece in one piece—Ohio Bulletin 68 says the fleece should be removed in one piece as nearly as possible so that it will be easier to pick up and tie. Avoid making second cuts, as the short fibers (noils) will comb out and, therefore, lower the value of the fleece.

Fig. 4.8—A properly shorn sheep. Notice absence of nicks and cuts as well as close, even clipping.

Chemical Shearing a Possibility

Scientists at the USDA Beltsville (Maryland) Research Center and many other places are testing a new drug that enables the sheepman to remove the wool from sheep easily, simply by pulling it off. It is caused by giving the sheep a new drug which causes a weak spot in the wool growth so the fibers break off. The name of the drug is cyclophosphamide. It can prove lethal in large doses (90 mg/kg); however, researchers indicate that doses in the 10 to 30 mg/kg body weight showed no signs of ill effects or toxicity—for example, 11.36 mg/lb. or 25 mg/kg. (See frontispiece illustration.)

It must be indicated that the drug is **not yet approved** for general use. The FDA will not release the drug for commercial use until it has been proven there is no residual effect on the meat. Nevertheless, for small producers it holds great promise, as sheep shearers are very expensive to employ and next to impossible to get for a small flock. Another use would be with lambs when they are covered with stickers and the wool is too short to shear; with this drug, the animals could be defleeced easily in order to promote health and well being. In addition, cuts and injuries due to shearing would be eliminated.

However, with large flocks, management presents a serious problem, as sheep must be protected for about 3 weeks from cold or sunburn after defleecing until enough wool grows back to shield them from the elements. Range flocks would have to be confined so wool could be collected, and as dosage is critical, individual treatment may be necessary.

The chemical can be given orally or intravenously, but relationship to body weight is important. The chemical interrupts cell growth in the bulb of each wool fiber, causing a ring-like weak spot. This weak spot moves up from the base as the fiber grows, reaching the skin surface in six to seven days. Then, the entire fleece can be pulled off easily without discomfort to the sheep, or any waste of wool. Not all sheep respond the same, so they must be regularly checked.

In spite of all the uncertainties, chemical shearing holds great possibilities, especially for the small pro-

Fig. 4.9—This is the proper method of tying a fleece. Fleece is first turned flesh-side down and tied in such a fashion to show the greatest amount of clean body wool.

ducers, and its development should be carefully watched.

Tie fleeces properly—All fleeces should be tied separately. Dead wool should not be tied up inside a "live" fleece. After shearing, it is a good idea to let fleeces air out a day or two before packing so that animal heat is lost; however, this is not a general practice.

Turn fleeces flesh-side down. Then turn the neck back to the shoulder, turn the britch and shanks up to the point of the hips, and turn the belly in until the

HANDLING SHEEP AND LAMBS

better wool shows. The fleece should be rolled in such a way as to show the greatest amount of clean body wool and tied into a neat package.

Use paper twine—Only paper twine should be used for tying. Wool tied with jute or sisal twines are generally subject to discount because the fibers from them become mixed with the wool and in subsequent processing do not absorb the dye the same as wool. Two strands each way are sufficient to hold even large fleeces. One length of twine eight and one-half to nine feet long is enough to tie the average fleece. One pound of single ply paper twine will tie 25-30 fleeces.

Do not tie too tightly—Cramming large fleeces into a small package may cause the grader to estimate its shrinkage too high since the bulk is not there for the weight of the fleece. It is a good procedure to turn the wool sack seam-side out and shake thoroughly before filling with wool.

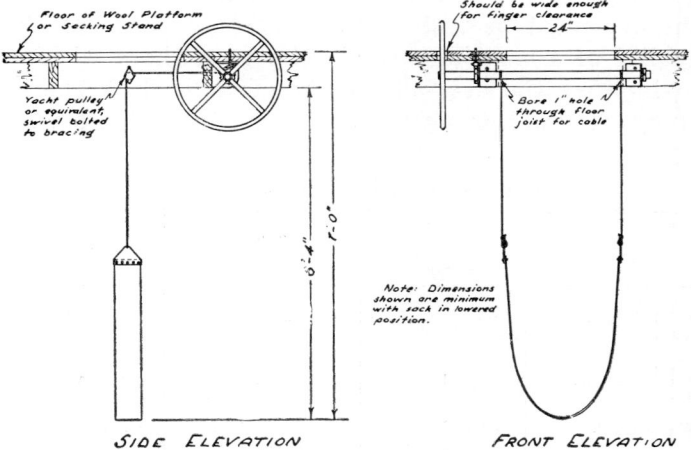

Fig. 4.10—Wool sacking windlass—an excellent kind of wool sacker. This device eliminates one man below to raise the sack for sewing. Complete plans are available from Agricultural Publications, Berkeley, Calif. Ask for Plan No. 73.

Sack tags separately—The tagging of sheep before actual shearing begins usually results in a higher quality clip. In any event, all heavy tags, sweat locks, and short leg wools should be kept out of the main fleece. Growers often have different ideas as to what constitute tags. Some virtually skirt the fleece, but all too many do not remove enough of the tags to improve the grade of their wool. Tags and sweat locks should be sold separately, as they lower the value of the entire fleece.

2. Grading Wool

Grading wool is really a job for a specialist, as it takes years of training and experience. The reason wool grading is different is that it must be graded by eye. Therefore, a great deal of experience, plus a certain amount of native ability, is necessary in order to properly do the operation. Wool fibers vary tremendously in diameter so that it is relatively impossible to get an average sample with any significance; hence, the eye must be trained to "grade" at a glance the quality of a sample of wool. However, those wishing to improve the over-all quality of their flock should study the various grades and learn to evaluate wool

UNITED STATES GRADES OF WOOL

Spinning-count system	Corresponding older American system	Spinning-count system	Corresponding older American system
80's 70's 64's	Fine	46's	Low ¼-blood
62's* 60's 58's	½-blood	44's	Common
56's	⅜-blood	40's 30's	Braid
50's 48's	¼-blood		

*A grade widely used in trade, but not yet recognized officially.

and wool clips to the best of their ability. A useful publication on wool grading is **Farmers' Bulletin 1805** obtained from the U. S. Department of Agriculture, Washington, D. C., and **Circular 171**, University of California, "California Wool Production." Additional information on wool grading is found in the chapter on marketing.

Separate wool within broad limits—Whenever possible fleeces of different grades should be separated and sacked separately. While not practical with small flocks, the wool from ewes, rams, yearlings, black fleeces, etc., should be sacked separately. Some operators make it a practice of separating a large ram fleece into two "ewe" fleeces because ram fleeces may be unjustly discriminated against.

Spinning counts—In the United States the practice of indicating the spinning counts as well as giving the American grade name of the wool is becoming increasingly prevalent. It permits a more accurate description of the fineness in diameter of the fiber. **The spinning count is based on the number of hanks of yarn (each hank 560 yards in length) that can be spun from one pound of wool top of a certain diameter, provided it is spun to its maximum fineness.** In actual practice wools are seldom spun to their highest count because of increased cost.

3. Examining the Fleece

Many times during various operations of handling sheep, growers find it necessary to look at the fleece. While outside appearance has some importance, the length, crimp and density can only be observed by parting the fleece. It is imperative that approved practices be employed while examining the fleece.

Never grab wool to hold sheep—Bruised spots appear under the skin wherever sheep are held by grabbing the wool. Sheep also tend to struggle more because of the pain caused by pulling on their wool. Therefore, catch and hold sheep by the body, head or neck. With a little practice in using one's hands and legs, sheep can be held quietly and with little effort.

Keep fingers together—This is advisable because there is less chance of running fingers promiscuously into the wool. Pressure is also exerted against a larger area so that no strain is put on individual wool fibers or small skin areas.

Examine wool by reflected light—A considerable

Fig. 4.11—This is a proper way to hold a sheep. One hand of the operator is under the chin of the sheep and the other is held in back of the head. Notice that the feet of the man are spread so he cannot be thrown off balance, with one knee against the chest of the animal. In this fashion, hands and body coordinate in restraining the animal. Never catch or hold sheep by the wool.

amount of light is necessary in order to properly examine wool so that crimp and length of fiber may be seen easily. However, the best light for examining wool is by reflected light, such as is found on the north side of a building. If one is outside in the open, it is a desirable practice to place the sheep so the operator can examine the fleece by bending over the animal with his back to the sun.

Open at natural breaks—With a little care it is possible to open the fleece of a sheep almost anywhere one desires and still take advantage of the natural cleavage of the wool so the fleece is not torn apart too much. Indiscriminate opening of the fleece causes it to tend to hang open so that foreign matter and dirt enter easily.

Use sides of palms—It is possible to open a fleece correctly by using one's finger; however, most sheepmen prefer to open the fleece using the side of the little finger and end of the palm of the hand. This gives a longer surface so that the wool is opened wide enough for a good view.

Close fleece when finished—After opening and examining the wool, it should be closed or pushed together so the entire fleece tends to hang together as one piece. The fleece should be opened just as little as necessary.

4. Utilizing the Gregariousness of Sheep

Sheep are gregarious in nature. Handling sheep can be made much easier if a breeder will study their gregarious habits, as well as other habits, and incorporate his knowledge into everyday practices. For example, it is almost impossible to separate one or two sheep from a band without first getting the entire band into an enclosure, because they tend to run as a bunch.

Flocking instinct in fine wool breeds—The Merino and Rambouillet breeds and their grades have the flock-

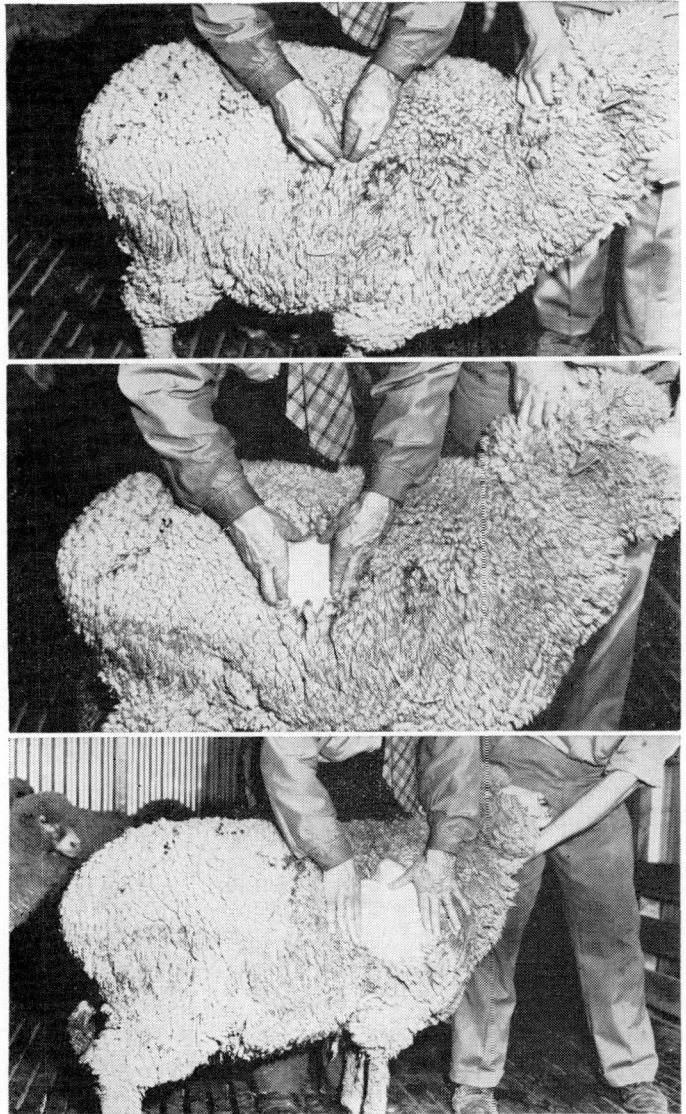

Fig. 4.12—These three pictures illustrate the proper technique in examining a fleece. Both hands with fingers together are first inserted at a natural opening in the fleece, then the flat of the hand is used to lay the fleece aside for several inches, thus exposing a large area to the examiner. (Courtesy, Prof. J. F. Wilson, University of California, Davis)

ing instinct developed to the greatest degree, while Cheviots and Black-faced Highland breeds are least gregarious. For this reason, range-herded sheep must have some fine wool blood, whereas farm flocks in fenced areas can be any breed.

Provide high, dry areas—Sheep prefer high and dry ground and in favorable weather prefer to be outside in the open. The younger sheep seem to like higher ground than older members of the flock. Young lambs often like to climb onto the top of a bale of straw or even onto the backs of their mothers.

5. Winter Handling

Winter management has an important relation to the returns expected from the flock. Native feed is often scarce, weather, of course, is generally severe, and as ewes are carrying lambs, it is quite important that special care be taken during the winter months.

Provide ample exercise—A moderate amount of exercise is necessary if the ewes are to produce healthy lambs. In cold weather, sheep have a tendency to bunch up and become inactive. If winter pastures are used, no other exercise arrangements are necessary. However, sheep in feedlots will gain exercise if their roughage is scattered over a field away from their shelter.

Keep away from the mud—Pregnant ewes especially should not be forced to wade through mud or deep snow. Sheep can stand severe cold, but once they become soaked, trouble begins. Guard against narrow doors, high sills, or ewes being chased by dogs, as loss of lambs and ewes may result. This is a real danger in winter, as sheep are continually coming into contact with buildings and shelters.

Keep fleeces dry—Dry snow has no ill effect as ewes readily shake it off. On the other hand, if fleeces become wet from rain or wet snow, colds and pneumonia often result. Rain never helps sheep.

Choose proven winter ranges—Range sheepmen often allow their stock to graze the entire winter season. This is commendable provided the range is known to be capable of supporting sheep during this season. Snow may provide enough moisture for drinking in some areas that are too dry during spring and summer months and, therefore, a certain amount of residual feed is available on such ranges.

Keep emergency feed—A good winter range has abundant feed, ample water, and shelter during stormy

Fig. 4.13—A good supply of feed is insurance against major losses. This simple shed is filled with baled and chopped hay so that ample tonnage can be stored in a small space.

periods. However, most successful sheepmen prefer to have some emergency feed available so that major losses will not occur in the event of sudden, severe weather. Even a straw stack is a welcome reserve feed supply during unusual years.

Go into winter in good flesh—Sheep that are in good flesh by winter are healthier, can stand the rigors of winter, and come out in better shape in the spring than sheep that go into the winter period in poor flesh. In addition, well-fleshed sheep are usually stronger and can work harder and find more feed when feed is scarce than those in less condition.

Feed poorest feed first—Alfalfa hay is the standard winter roughage for sheep in feedlots although native hay is fed when alfalfa is unavailable. In any case, coarse, stemmy feed should be fed first as sheep will not be as choosy when they first come off the range. Later on, the better hay can be fed after sheep have cleaned up the poorer hay and are also in need of better feed themselves. Three to three and one-half pounds per day of good hay will nicely winter ewes up to 150 pounds. Succulent feeds, such as beet pulp, roots, and silage, make excellent winter feeds.

Many large operators like to provide their ewes with one-fourth pound daily of cottonseed meal or cake while grazing on winter range in order to insure good health and prevent loss in body weight.

6. Setting Up Sheep

For one reason or another, sheep, particularly farm flocks, are continually being handled. When trimming feet, examining heads for foxtails, checking udders, etc., it is necessary to catch and hold sheep. Proper handling in such cases is not only easier on the sheep but on the operator as well.

Handle carefully—This may seem like superfluous advice; however, sheep are not strong animals compared to other domestic animals and are easily hurt. Bones can even be broken by rough handling. In addition, fingers should be kept together and not poked into the wool or flesh. Under no circumstances should sheep be held or grasped by their wool.

Confine flock first—It is very difficult to catch a single sheep out of a large band. Therefore, if the entire flock or at least a sizeable number are first driven

Fig. 4.14—Sheep should be driven into an enclosure and confined if they are to be caught easily.

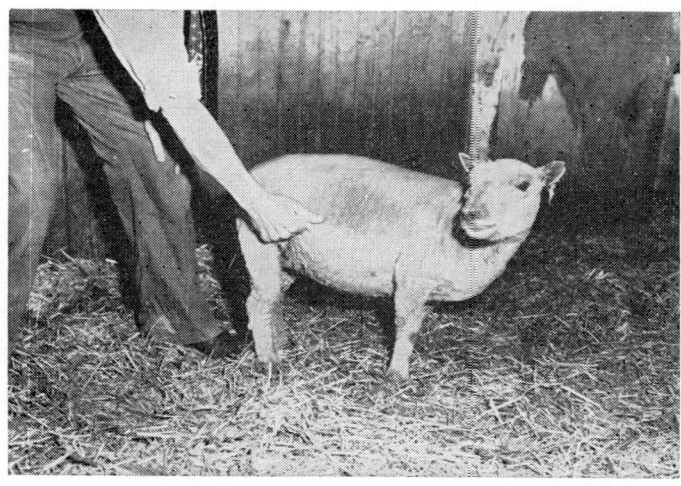

Fig. 4.15—An excellent way of catching and holding a sheep. Sheep caught and held by the rear flank, as illustrated here, seldom attempt to struggle.

HANDLING SHEEP AND LAMBS

into an enclosure and are temporarily crowded, it is comparatively simple to walk into the band and pick out the desired animal.

Catch by rear flank—The flank hold is one of the best for catching and holding sheep (see illustration), as even large sheep can be held in this fashion. In addition to the hand, the operator also uses his opposite knee to steady the animal. Sheep may also be held by catching a hind leg or by holding it around the neck.

Use dogs only on bands accustomed to dogs—A good-working dog is a real asset. However, the sight of a strange dog to sheep unaccustomed to being worked by a dog, may cause them to become frightened and run. Many fat sheep and ewes heavy with lamb die of excitement and overexertion when running.

Fig. 4.16—A good sheep dog is worth two or three men. Here is Bob Finlay, herdsman, University of California, and his top-notch Border Collie, Bill.

Set up on rump—In order to examine sheep or trim their feet, they should first be set up on their rump. One approved method is to stand on the left side of the sheep with the left hand under the jaw and the right hand grasping the left hind leg just above the hock.

Fig. 4.17—This is one method of setting up a sheep. The sheep's head is forced back by pressure of the hand under the chin as illustrated, while lifting on the flank, not the wool. Upsetting is accomplished by rolling the sheep over the operator's knee. Both knees of the operator are bent, but the right knee is bent more so the sheep rolls over it when upsetting.

The operator should pull the leg forward and up, at the same time lifting up and back with the hand under the jaw. This causes the sheep to sit down on its rump directly in front of the operator and fall back against his knees.

7. Marking Sheep

The term "marking sheep" refers to the yearly task of docking and castrating lambs. Both of these operations are essential for healthy sheep and to produce desirable meat carcasses. Dung, urine, weeds, etc., collect on the tails of undocked sheep which serve as potential screw worm and maggot trouble spots. Buyers will always discount undocked or uncastrated animals. According to **Cornell Bulletin** 828, long-tailed rams are discounted on the average of $1 per 100 pounds. On

HANDLING SHEEP AND LAMBS

such a necessary operation, it is important to use correct methods in performing the required skills.

Mark sheep young—It is best to dock and castrate lambs when they are from one to three weeks of age. Regardless of the method used, there is less setback in growth if these operations are performed when the animals are young. With large bands, from a management standpoint, it may be necessary to let many of the lambs get to the three week stage or slightly older; however, even extensive operators prefer to mark lambs young and, therefore, make it a practice to have a man with the sheep continually to assist at parturition and shortly thereafter to mark young lambs.

Use modern methods—Until recently, almost all lambs were docked with a knife or hot iron and castrated in this fashion, first cutting the lower third of

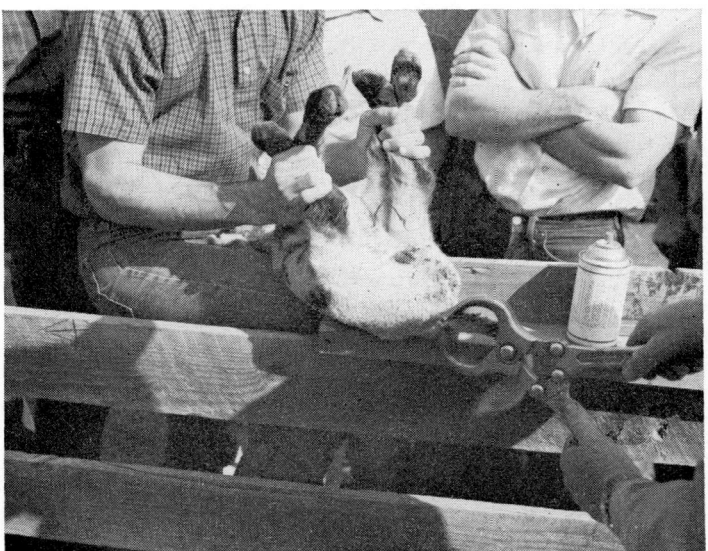

Fig. 4.18—Illustrating one way to hold a lamb when using the Burdizzo for docking.

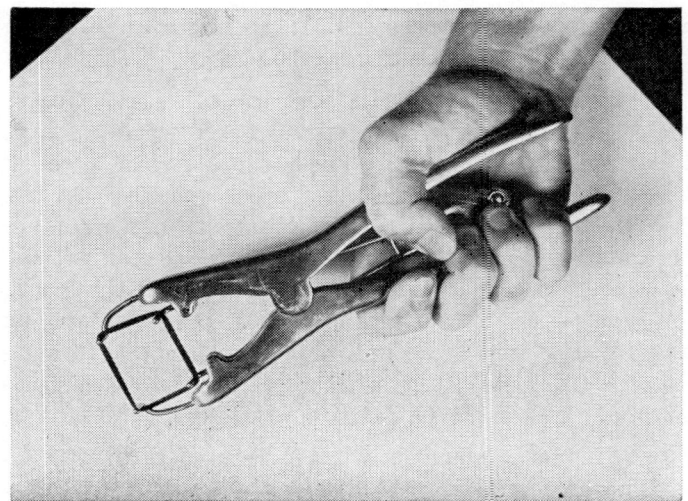

Fig. 4.19—The elastrator is a modern docking and castrating instrument.

the scrotal sack off, then pulling the testicles and cords out either with hands or teeth. While this method was satisfactory, it was a messy job and dangerous during fly-strike time.

Since the development of the elastrator, the operation has been speeded up in addition to eliminating the danger of fly infestation, because the method is bloodless, plus the fact that a messy job for the operator is eliminated. It is important to follow manufacturer's directions carefully when using the elastrator. In addition, ordinary sanitary precautions and careful observation and follow up of marked lambs should be practiced.

Docking consists of placing a rubber band around the tail about an inch from the body. In about two weeks the tail will drop off.

Castration is similar in that the same type band is placed over the scrotum and above the testicles which soon atrophy and drop off. After the first discomfort, the lambs pay little attention to the rubber band and continue to eat as before. Many shepherds carry the instrument near or with them and mark a few lambs of the right age every day during lambing season. Some shepherds vary the length of the tails as a means of identification. For example, ewe lambs may be docked shorter than ram lambs.

Choose warm weather—Bright, sunny days are the best times to mark sheep as they do not have to withstand the shock of operation and weather, too. This is particularly advisable if sheep are docked and castrated with a knife.

Fig. 4.20—The tail of this lamb is about to drop off as a result of an elastrator band placed there two to three weeks previously. This is a clean, easy way to dock lambs.

Mark on clean ground—Temporary pens in an unused corner of a pasture are superior to dusty corrals. Navel ill, stiff joints, and general infection are always more apt to occur when marking takes place in dirty areas where sheep have been known to congregate.

Mark before fly time—By using the elastrator, this precaution is not as imperative as with other methods. Also, in some climates it cannot be avoided. However, whenever possible, it is a good practice to follow as it eliminates one hazard.

Treat cuts—A good disinfectant like iodine plus a fly repellent (screw worm smear) should be applied to any break in the skin or flesh.

Keep on dry, clean pasture—After marking, lambs and ewes should be kept on a clean, parasite-free pasture. Irrigated pastures are good, but lambs should

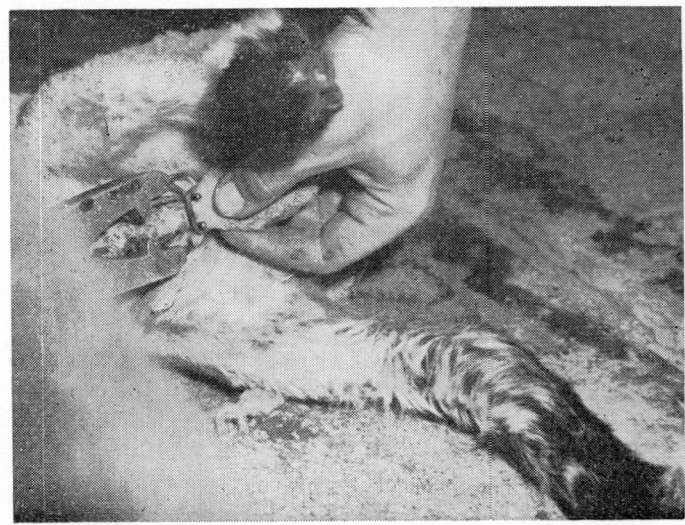

Fig. 4.21—Proper placement of the elastrator band is essential for a good job of castration. The skin is unbroken by this method.

HANDLING SHEEP AND LAMBS

Fig. 4.22—Put ewes and lambs on clean pasture after marking.

be kept away from boggy or swampy areas until all wounds are well healed. The pasture should also be reasonably high so when animals lie down they do not touch the end of the tail to bare ground and thus increase the possibility of tetanus.

8. Tagging Sheep

Stained wool around the vent and wool matted with manure should be removed from sheep. Such wool when removed is termed tags. Tags stain, and discolored wool, if left on the animals, interferes with normal lambing activities. Tagging is really a constant operation, as tags should be removed whenever they become a source of irritation although certain times of the year are more or less noted as tagging seasons.

Watch confined sheep—Sheep penned up are more apt to need tagging than range sheep. Sheep may require more attention during damp weather than during dry weather. It is best to tag with power shears although small flocks can be done quickly with a hand shear.

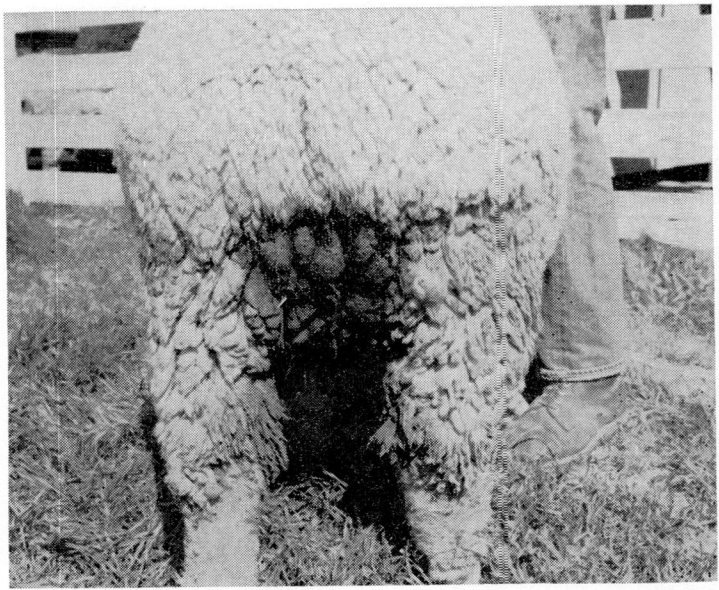

Fig. 4.23—A sheep badly in need of tagging. Notice the dung locks and stained wool around the rear end that must be removed. Lambs have difficulty finding the teats when wool grows around the udder.

Tag before shearing—Before starting to shear, all tags should be removed from the sheep and placed in a separate pile. Wool buyers discount the value of fleeces if tags are left in with the fleece. Tagging during shearing eliminates catching the sheep specifically for tagging, and is a desirable management practice.

Tag ewes prior to lambing—Wool should be shorn from the vent and around the dock a few weeks before lambing. This eliminates a large number of tags being formed because of unusual amounts of watery discharges during and prior to lambing. **Cornell Extension Bulletin 828** says loose wool from dock, sides, and backs should be removed, and for many ewes it may be advisable to shear around the udders, especially near the

HANDLING SHEEP AND LAMBS

teats. Long locks of wool here will oftentimes be sucked by the lamb, mistaking them for the teats. Long wool around the vent of ewes may occasionally have to be trimmed away to aid breeding.

Store tags separately—Tags will spoil an entire bag of wool, so they should not be sacked together. Roll tags up, skin side out, and put in sacks. According to **Farmers' Bulletin** 840, wool buyers do not like tags stored in wool boxes.

9. Trimming Feet

Unless the feet are trimmed properly, bad posture and feet often result. Rams in particular may become shy breeders if bad feet prevent them from getting around to all ewes in heat. Well-trimmed feet reduce the danger of foot rot.

Fig. 4.24—An instructor in animal husbandry teaching students to tag ewes.

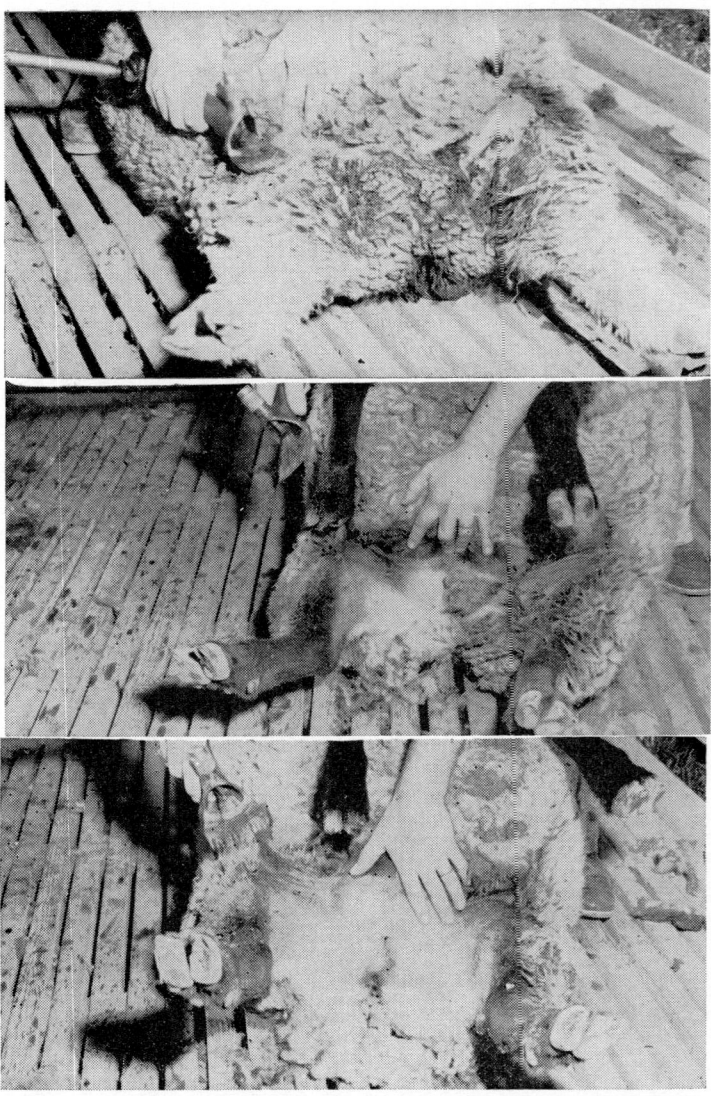

Fig. 4.25—Three stages in proper sequence of tagging a ewe. Top picture shows how ewe is properly set up on rump and held against knees of shearer. Middle picture: Note where shearer starts to trim on inside of upper part of leg. Bottom picture shows how shearer's hand protects udder as he removes wool around vent.

Fig. 4.26—Here is the proper way to set up and hold a sheep in order to trim its feet. Hoof trimming is an approved practice all sheepmen should master. Either a sharp knife or hoof trimmer can be used. Feet should be inspected often and regularly, especially if animals are on irrigated pastures or soft ground. Overgrown hooves can lead to many troubles in addition to keeping the sheep from grazing actively.

Trim regularly—While hoofs should be trimmed whenever necessary, periodic inspection throughout the year insures every sheep getting trimmed when needed. Once per year should be the minimum time, and breeding sheep on wet or irrigated pastures may have to be inspected two or three times. Range sheep on hard ground may seldom if ever need trimming but should still be watched carefully. Some producers consider it a good procedure to trim all feet at shearing time.

Use trimming shears—If a large amount of trimming is necessary, a sturdy, foot-trimming shear is the best instrument as it will speed up the operation considerably. An ordinary pruning shear or pocket knife is also satisfactory for occasional use. A rasp is handy to shape up the foot after trimming, although not always necessary.

Fig. 4.27—A hoof properly trimmed. Note how short the hoof may be trimmed. Occasionally blood is drawn, but this is seldom serious.

Set up on rump—The correct position for trimming is to set the animal up on its rump so it will not struggle (see information on setting up). Ordinarily, trimming is a one-man operation.

Wet tough hoofs—If hoofs are trimmed during the rainy season, they will cut more easily. Tough hoofs can be softened by driving the band into a marsh area for several hours.

Square up hoofs—Care should be taken not to cut too deep or bleeding and soreness may follow. With especially long hoofs, greatly in need of care, it may be desirable to do the job in several stages a few weeks apart. Hoofs should be trimmed so that sheep appear to stand squarely on their feet with legs straight. Well-trimmed hoofs give animals the appearance of standing neatly on their toes.

10. Painting Rams and Branding Sheep

It is necessary to brand sheep for various reasons, especially in the western states. One reason is to show ownership or identification. Another common reason for branding is to identify ewes and lambs at lambing time or to show the "buck band" during breeding season. Improper branding or use of materials can cause great damage to the wool clip so that approved practices must be followed.

Use only specially prepared branding fluid—House paint, tar, and other such compounds are insoluble in scouring and cannot be taken out by ordinary scouring methods. Another reason for using a prepared fluid is that brands are generally placed on the best wool so that incorrect fluid ruins the highest grade wool. Good sheep branding liquids should stay on the sheep for a year, yet wash out in the normal scouring process.

Brand as small an area as possible—While brands

must be conspicuous and easily read, there is no point in using unnecessarily large brands.

Brand after shearing—Common practice is to brand immediately after shearing as sheep are already confined and, of course, the old brand has been removed with the fleece.

Prepare your own ram-marking paste—Ordinary lubricating oil mixed with yellow ochre, venetian red, lamp black, or other coloring material, until it forms a paste, is desirable for painting the ram's chest during the breeding season. Never use linseed oil, as it will harden. An old two-inch paint brush is excellent for applying paste.

Paint chest every day—During the breeding season, a fresh supply of indicator paste (same color) should be smeared on the ram's chest every day or so in order that the ewes he is mounting will be properly marked. On the 18th day the color should be changed so it can be determined which ewes are returning in heat.

11. Telling Age of Sheep

In the absence of definite records as to year of birth, the front teeth are the most reliable index of age until the sheep is four years old. After that, it is more difficult and depends to a great extent upon the type of soil and vegetation being grazed.

Open mouths with fingers—Everyone engaged in the sheep business should learn to examine a sheep's mouth properly. Mouths can best be opened by using the index and middle fingers to part the front lips as illustrated.

Learn to tell age by teeth—The lamb has 20 temporary or milk teeth. They are smaller and somewhat darker than the permanent teeth, but for telling age, only the front teeth in the lower jaw are used. At from 10 to 14 months of age the two center temporary teeth

HANDLING SHEEP AND LAMBS

are replaced by two broad permanent incisors. The second pair on either side of the yearling teeth come in at 22 to 26 months. Three-year-old teeth come in at 34 to 37 months, and a full mouth (8 cutting teeth) at

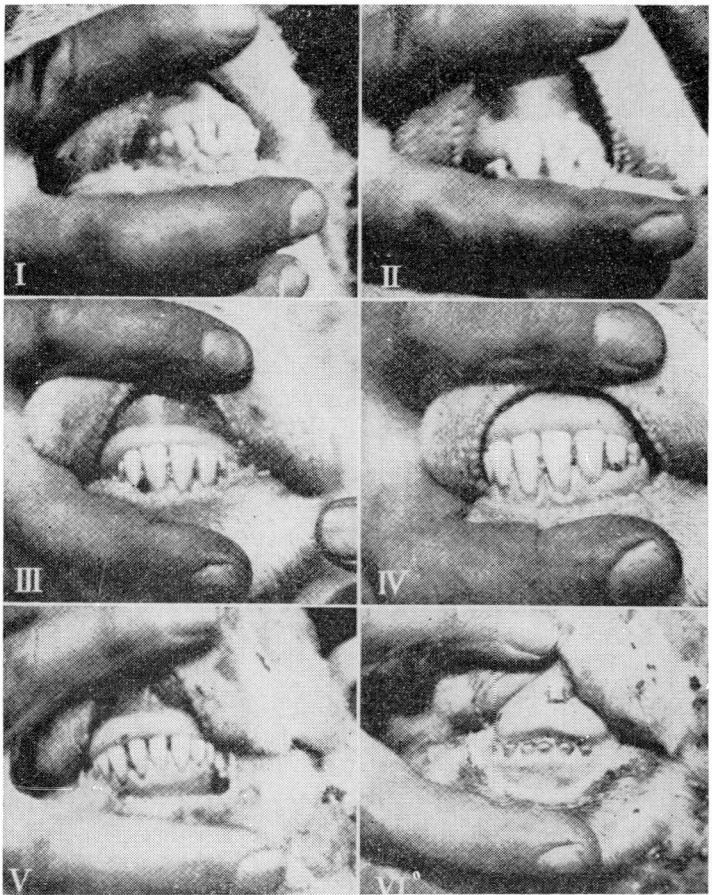

Fig. 4.28—Photographs showing teeth of sheep. Top left, a lamb's teeth; top right, teeth of yearling; center left, teeth of two-year-old. Center right, teeth of three-year-old. Bottom left, teeth of four-year-old (full mouth); bottom right, an old sheep's badly worn teeth.

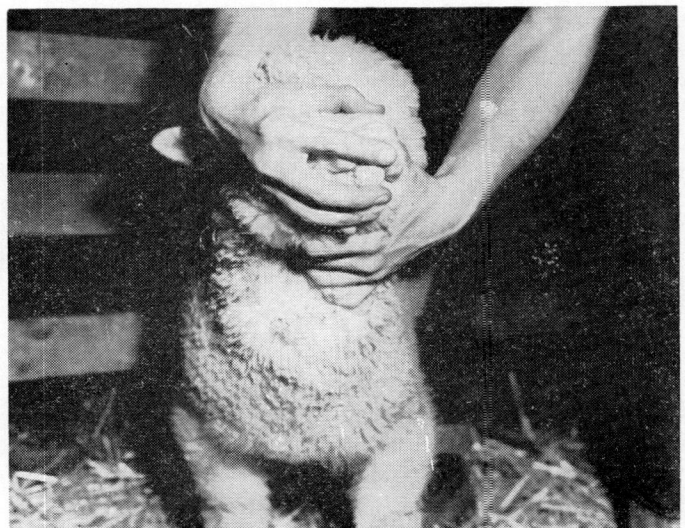

Fig. 4.29—Use your fingers, as illustrated, to open a sheep's mouth so that teeth may be examined.

around 48 months. After that, age cannot be told as accurately, but the tendency is for the teeth to become shorter and wider apart as the sheep gets older. When a sheep loses one or more permanent teeth, it is called "broken mouthed." If all the teeth (incisors) are gone, the sheep is known as a "gummer." (See Figure 14.4, Chapter XIV.)

Find out where sheep originate—There is some breed difference regarding age at which teeth come in; however, the condition of the teeth is an indication of maturity rather than the exact age of the animal. The age at which this condition appears depends to a great extent upon the kind of grazing land on which the sheep are kept. Many sheep have broken mouths at six years of age when forced to graze on sandy soils with short, sparse vegetation. Those grown on heavy soils with good pasture growth may have sound mouths at seven or eight years of age. As long as a ewe has a sound mouth, she will remain in good health and would not be classified as a cull. On the other hand, it is advantageous to know the history of ewes when purchasing them as they may appear better than they really are if grown under unfavorable pasture conditions.

12. Tattooing Sheep

In order to identify animals positively, sheep can be tattooed. The practice is not as common as with other classes of livestock such as dairy cattle, but some breeders do utilize this method. Light skinned breeds are easiest to mark; however, a reasonable degree of success can be obtained with dark skinned breeds like Hampshires or Suffolks. The operation can be done on any age sheep. Regular tattooing ink is best. Green or red ink has been used on dark skinned breeds, but results indicate little improvement over black ink.

Tattoo inside of ear—This is the best location as less hair is found in this area. Care should be exercised so marks are made between the upraised ribs of the ear.

Clean well—Alcohol is best, although soap and water or a clean, dry rag will help. This is an important step as the clearness of the mark depends on having the ear absolutely clean and free of grease.

Clamp down hard—The instrument used to make the identification must pierce the skin and draw blood.

Rub in well—After tattooing, the ink is applied and must be rubbed well into the punctures in the skin. No further care is needed.

13. Training Show Sheep

Sheep that are to be shown are handled in many respects like other sheep on the farm except that extra care and attention must be employed. They should be

Fig. 4.30—The picture illustrates the proper procedure in learning to train show sheep. Sheep should not be photographed against a wire fence background.

especially fed for this purpose and have the fleece and feet in proper condition for critical inspection.

Protect animals from weather—It is important that the animals be protected from hot sun, rain, and flies. During summer months in most parts of the country, they should be kept in the barn in a well-ventilated pen bedded with clean straw. Do not let sheep out in the rain, as it washes out the natural oils and gives a dry, harsh appearance. Protect animals from dust and foreign matter such as burrs and foxtail.

Dip sheep after shearing—After sheep have been shorn in the spring, it is advisable to dip them shortly afterwards. Not only are external parasites controlled, but the skin is cleaned and the wool takes on a brighter, fresh appearance.

Wash mutton breeds early—This is not a must, but should any of the mutton breeds become too dirty, it is advisable to wash them before showing. It is important, though, that this be done a month or six weeks before showing so that the oil will return to the fleece.

Drench infected sheep—It is almost impossible to get severely infected sheep into showing bloom without drenching them. Therefore, if internal parasites are a problem, it is good practice to drench them. (See Chapter VIII.)

Fig. 4.31—Members of the FFA (Minster, Ohio) dipping sheep as a community service project. They are using a portable dipping vat.

Blanket fat lambs—A clean fleece is very desirable when showing fat lambs; therefore, they should be blanketed in order to keep out dirt and grease. Blankets can be made from old grain sacks. Use good sacks and slit them down the seam and in the corner of the sack bottom. Cut out a hole opposite the seam side for the lamb's head. Two small holes cut in the edges of the sack near the rear flank can be made, through which a strip of burlap can be inserted to tie loosely around the hind legs.

Block fat lambs early—A month or so before show time the lamb should be blocked for the first time. Do

Fig. 4.32—4-H members blocking a lamb for the Junior Livestock Show at the Cow Palace, San Francisco, California. Blocking should begin at least a month before show time. Modern shows do not require as much wool or blocking as previously was the case.

not remove any wool in the twist, but trim the back down flat to within a fourth-inch or so of the backbone. This is particularly true of wethers. Final blocking can

HANDLING SHEEP AND LAMBS

be done a little at a time by going over the fleece several times before show time. Lambs should appear square without giving the appearance of being trimmed.

Train show sheep to stand—Sheep should stand quietly in position, particularly while being examined. This cannot be done with one operation, but is relatively easy if the animals are worked with a little every day or so. Do not work with them too long at one time—at first only three or four minutes—so they do not become tired. Train them to be alert at all times and to stand squarely on their feet. Do not lean on the lamb but hold him by the cheeks or under the chin.

Fig. 4.33—A 4-H district show in Texas. If lambs are to behave at a show, they must be trained early to stand quietly.

14. Diagnosing Pregnancy

Of considerable interest to producers is the work on early diagnosis of pregnancy in sheep.

Dr. Lavon Koger, DVM, of Washington State University, reported that it was possible to use rectal palpation for pregnancy diagnosis which should be accurate at about the thirtieth day of pregnancy, and that the operator would be able to tell if an embryo was present in one or both horns of the uterus, and also determine the number of embryos. The problem is finding someone with a small enough hand (should fit into a No. 6 glove) in order to perform the palpation. Women may be able to do this better than men because of small hands.

Dr. Koger has also been working on a method of digital palpation via the vagina to detect the tip of the cervix. In the nonpregnant ewe, this has a firm or hard feel, and the spiral folds feel cartilaginous. When the ewe is in season and after she has conceived, the cervix feels soft and yielding. At the fiftieth to sixtieth day of pregnancy, you cannot find the cervix, which has fallen under the rim of the pelvic girdle and is below that level in the abdomen.

Dr. Koger has been using the cervical palpation with a relatively small group of ewes. The method shows some promise, but does not appear to be as accurate as the rectal palpation method at this point.

Dr. Clarence Hulet of the U. S. Sheep Experiment Station at Dubois, Idaho, has developed a restraining device to hold the ewe on her back with the rear end up at a 45-degree angle. He shears a small area immediately in front of the udder and makes a small incision in order to insert one finger, with which he palpates for pregnancy. He does this very rapidly; for instance, he recently did 32 ewe lambs in one hour (these were 8 to 9 months of age and had been bred in conjunction with an estrus control trial). On some individual ewes, he has operated as many as six times to date, and has operated on as many as 224 yearling ewes in one day. He has done several thousand head all together, and he reports a high degree of accuracy in diagnosis. The incisions, which are made immediately to the right or left of the midline in front of the udder, are closed with a mattress suture and a continuous suture thread. The wound area is treated with nitrofurozone and, in addition to the suture, is held together with one or two skin clips. Dr. Hulet said that this method can be used to detect pregnancy as early as 30 days after breeding.

Dr. Hulet indicated that, aside from other considerations, the early pregnancy diagnosis is valuable in that it allows producers to cull non-breeding ewe lambs at an early age, where often those culled can be sold directly to slaughter outlets. At present, the experimental ewe lambs at Dubois culled under this program are going to market within 7 days after the operation.

Other reasons for use of early pregnancy diagnosis include the following:

1. To cull dry ewes early in order to hit an early and higher market where they are to be sent to slaughter.
2. To save the additional feed for the remainder of the season by marketing unbred culls earlier. In operations where space and high quality feed are at a premium, it may be desirable to use early pregnancy diagnosis in order to sort out the dries so they can be held on lower cost and low quality feed.
3. To permit purebred breeders or small flock operators wishing to flock breed to single rams to check the breeding performance as early as possible in order to insure a future lamb crop.

CHAPTER V

RAISING LAMBS

Ask any sheepman what he considers the most important season of the year regarding success in sheep raising and he will answer without hesitation, "the lambing season." More time, effort, and patience are required during this activity than during any other phase of the sheep enterprise. It has been said, "twenty-four hours a day is the price a sheepman must pay for his profits during lambing season." However, there are many other important factors to consider in addition to time and patience. For example, production unit per ewe is one of the yardsticks of profit. Twin lambs weighing a total of 160 pounds and selling for 30¢ per pound will yield more profit than a single hundred pound lamb selling for 35¢ per pound. Therefore saving every lamb and maintaining a high percentage lamb crop is a "must." In order to take advantage of this important phase of the sheep industry and thus create more profit, it is highly recommended that the sheep breeder follow approved practices during the lambing season.

Activities Which Involve Approved Practices

1. Handling bred ewes.

2. Examining udders.
3. Caring for ewes before parturition.
4. Care at birth.
5. Caring for newborn lambs.
6. Handling ewes after lambing.
7. Feeding young lambs.
8. Creep feeding.
9. Preventing diseases of lambs.
10. Weaning lambs.
11. Preventing prolapse of rectum.
12. Controlling predators.

Fig. 5.1—Twin lambs and their mother. A high percentage lamb crop is the key to profit in the sheep industry. Many twin lambs are necessary in order to get a high percentage lamb crop. Notice the lamb on the right has just lost its tail as a result of docking by use of the elastrator.

1. Handling Bred Ewes

The care of ewes during the last month of pregnancy will vary a great deal according to the type of sheep production being followed. Many range men will lamb on the range except in very cold climates, so these

RAISING LAMBS

sheep rarely see a barn. Some range men prefer sheds or a sheep barn for lambing. Farm flocks, on the other hand, commonly lamb in sheds. In fact, many sheepmen make it a practice to bring in the ewes at night two or three weeks before the lambing season starts. However, if weather is mild, lambing in the open on slightly rolling hills is a desirable practice.

Keep ewes away from other livestock—Many times horses or even cattle will "bother" pregnant ewes and occasionally even kick and "run" them. Ewes should be left as undisturbed as possible and allowed to be regular in their habits during this period. Dogs, other

Fig. 5.2—A top job of clipping around vent and udder in preparation for lambing. Wool has also been clipped away from face and eyes.

than shepherds' herd dogs, should not be allowed around the ewes. It is understood that only pregnant ewes make up the band at this time, as wethers, rams, etc., should be separated.

Clip wool—The wool around the udder, teats, and vent, if not already tagged, should be removed. Don't neglect the eyes, especially on some wool-faced breeds. Clip this eye wool away, also, so that the ewe can find and feed her lamb.

Feed well—All sheepmen agree that overweight ewes becomes sluggish and are apt to have difficulty during lambing. However, at no time should ewes be allowed to lose flesh during pregnancy. Iowa State Bulletin P-107 recommends that ewes be fed one-half to three-fourths pounds of grain in addition to good legume hay. Equal parts of shelled corn and whole oats is a satisfactory grain ration although a mixture of six parts whole oats, three parts shelled corn, and one part protein concentrate is better. The amount of grain to be fed will depend to a great extent upon the quality of pasture, as many sheepmen who feed high quality alfalfa hay or have good irrigated pasture feed little or no grain if the ewes are in good flesh. It is believed that good feeding will help prevent "pregnancy disease" or "before-lambing-paralysis."

Get equipment ready—Pens and equipment should be put in order **before** the lambs start to arrive. The best time to make repairs is after lambing season so that the equipment can be sterilized by sunshine and made ready for the following lambing season. In addition, the herder knows exactly what to fix and repair on each piece of equipment as he has used it recently.

2. Examining Udders

Every careful shepherd examines the ewe to see that teat canals are open and that the wool around the udder

RAISING LAMBS 143

does not hinder the lamb when nursing. In large range flocks, this is not a common practice unless trouble is expected.

Examine frequently—Not only should udders be examined when the lamb is born, but also for several days after birth to see that the lamb is taking all the milk. Ewes can be held with one hand if only the pliability of the udder is under question. In most cases it is advisable to "set up" the ewe in order to examine her properly. Udders should be warm and firm, but not feverish or lumpy.

Fig. 5.3—An excellent udder. This udder is large and pliable yet not lumpy or feverish.

Separate swollen udders—Ewes with swollen udders should be separated from the rest of the flock to prevent spread of possible infection. It is recommended that beginners examine many udders until they can recognize a dangerously swollen udder, as all udders are somewhat swollen at parturition. Such udders should be milked out by hand once or twice a day until they return to normal size and the lamb begins to take all of the milk. Infected udders will become caked if they are not milked out.

Treat sore teats—The sharp teeth of lambs may cause the teats to become sore. **Iowa State Bulletin P-107** says that these sores should be opened as soon as they are discovered and treated twice a day with tincture of iodine.

3. Caring for Ewes Before Parturition

Everything possible that can make the actual birth of a lamb easy and successful should be done early, as it may be too late after a ewe has paralysis or trouble begins to develop during parturition. No matter how conscientious a sheepman may be, he may not be on hand to render aid or the ewe may go unobserved when she really needs help. This is especially true in large bands, and for this reason producers should reduce these risks as much as possible by planning a careful program in advance.

Separate heavier ewes from late lambers—In small flocks this may be unnecessary as the shepherd can easily keep all sheep under observation. It is a common procedure in large bands to make a practice of having a "drop band" close to headquarters, so they can be watched easily. This is accomplished by every ten days running the entire band through a chute and separating the ewes about to lamb. Size of udder and tenseness at the base of the teat are indications of the approach of lambing. Those ewes just making bag and those who will lamb in about five or six days are marked different colors with a piece of chalk, then dodged into separate fields.

Watch first lambers—Young ewes with a small udder oftentimes will lamb before older ewes with large, soft udders. First lamb ewes may not care for their lambs as well as older, dependable ewes. In addition, they are more susceptible to lambing troubles. For these reasons, a careful watch should be kept on them.

Provide exercise—Confinement may produce lambing paralysis plus a lack of general tone and alertness. Ewes should not be driven long distances, but a mild amount of exercise is beneficial as it generally makes for easier lambing. Ewes on hilly pasture generally will get enough exercise.

Prevent lambing paralysis—There are many forms of paralysis but one malady particularly associated with ewes just prior to lambing is called "Pregnant Ewe Disease." **Ohio State Bulletin 68** states that many sheepmen complain of ewes becoming paralyzed in the rear quarters shortly before lambing. It is most prevalent in ewes that are losing weight or those carrying twins. When paralysis occurs, the **grain** in the ration should be immediately **increased.** Stricken animals may be relieved by administering four ounces of molasses diluted with warm water as a drench. Make sure plenty of clean water is available.

Shear ewes completely prior to lambing—A common and desirable practice is to tag ewes prior to lambing in order to keep them clean and so the lambs can find the teats and nurse properly. This is sometimes done immediately after lambing or more often a month or two before. In direct contrast to this, some breeders shear the entire ewe about six to eight weeks prior to lambing time. There are two major advantages in this latter practice:
1. If the operator has to handle the animals during rainy weather, he keeps relatively dry as compared to handling animals with a soggy fleece.
2. When ewes are shorn they will come into a shelter and not stand out in the rain, thereby

facilitating herding activity and, most important, dropping the lambs in a dry place.

If shorn, ewes are easier to handle and observe during lambing period. Even in cold climates the animals do not seem to suffer from weather providing they have an opportunity to keep dry. Under this practice, yearly fleece weights tend to be slightly less because of the absence of wool grease, but the increased number of lambs saved and easy handling greatly outweigh any weight differential. Lawrence Heringer, rancher of Fall River Mills, California, has used this practice on several hundred ewes for a number of years with complete success.

4. Care at Birth

This is the most crucial period; however, it is not one for alarm if adequate precautions have been taken and approved practices followed **before** parturition approaches.

Be on hand—Lambing is, of course, a normal part of the reproductive process. Therefore, the vast majority of ewes will lamb normally and with no assistance. However, the wise shepherd keeps an eye on his sheep, for under the best conditions some ewes may need help.

Learn to recognize lambing signs—Usually the ewe has a sunken appearance on each side of her rump and the udder becomes full and teats distended. She becomes restless, paws at the bedding, and attempts to make a bed. She may even turn her head toward her rear quarters and bleat for her lamb. It is a desirable practice for beginners to learn to recognize these symptoms so they can manage the ewes properly.

Pen early—If pens are to be used, especially when weather is bad, they should be used early. Generally, one or two days in advance is enough, or as soon as

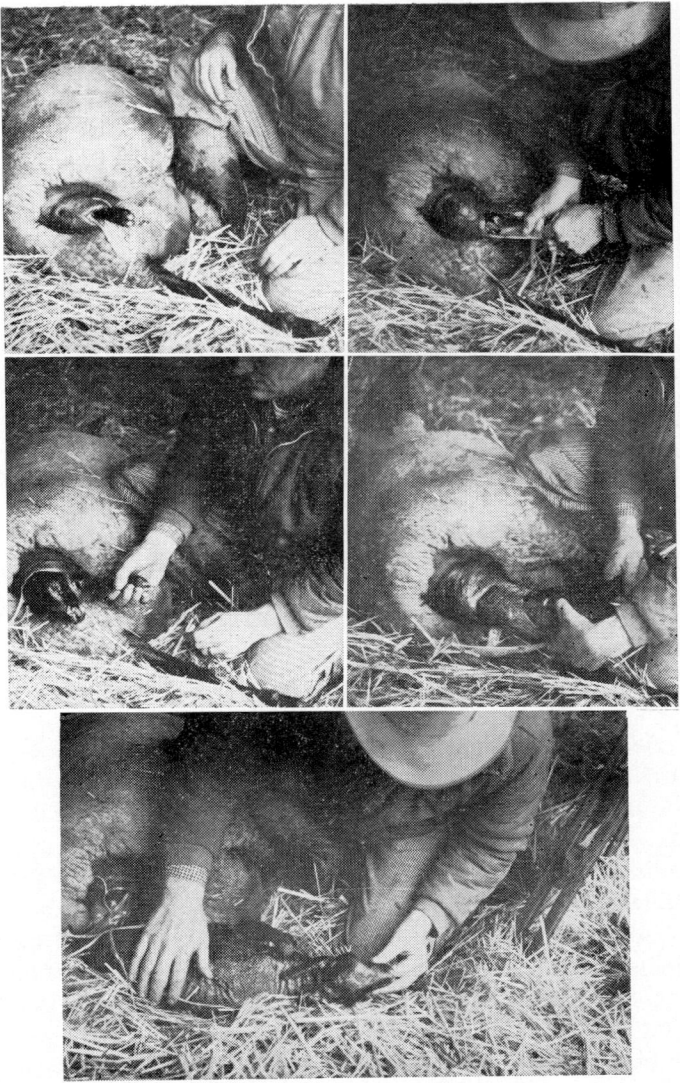

Fig. 5.4—These five pictures show the birth of a normal lamb and proper assistance given by the herdsman. Note that front feet come first and that should the herdsman assist, he pulls down toward the hocks of the ewe. In the last picture, he is cleaning mucus from the mouth of the lamb so it can breathe properly.

ewes become restless and show signs of lambing. In mild climates, many sheepmen do not use pens at all until trouble develops and they wish to confine the ewe and lamb.

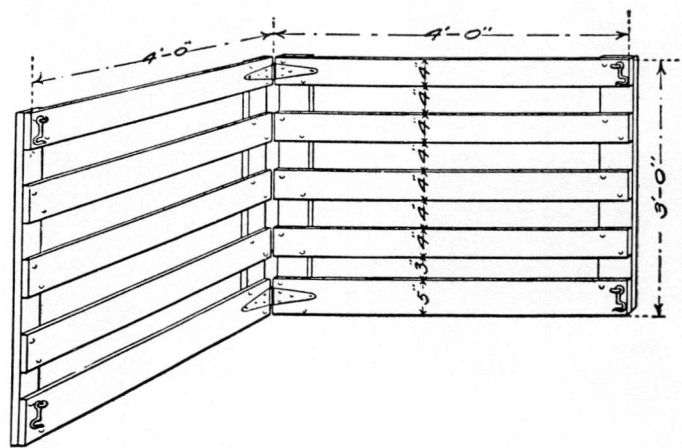

Fig. 5.5—Hinged panels for temporary lambing or claiming pens.

Assist only when necessary—Do not disturb the ewe during the first stages of labor. Allow her to produce the lamb herself if she can. If the water bag is broken and she does not deliver the lamb in a few minutes, assistance must be given. According to **Iowa State Bulletin P-107,** the hands and arms should be scrubbed in soap and water and the fingernails trimmed and cleaned. Bathe the hands and arms in a mild disinfectant, and then grease them with thin vaseline or a good mild soap lather to make entrance easier. Care must be taken to avoid injury to the ewe. It would be better for a person with small hands to make the examination.

Place the ewe in a corner and have an assistant elevate the hind quarters. This procedure will allow the lamb to fall back into the uterus and make the entrance easier. If you are alone and expect to use your right hand to enter the ewe, the ewe should be laid on her right side as carefully as possible.

In a normal presentation the lamb has the front legs extended and the head between or resting on them. Do not attempt to deliver the lamb in any other position until the presentation is corrected. Occasionally a young ewe will need assistance even with a normal presentation, because the lamb's head may be too large. Attempt to get the front legs out, then pull steadily on the legs and press in on the vulva just back of the lamb's head. Pull only when the ewe labors. After the head and feet are delivered, pull the front legs outward and down toward the udder to deliver the shoulders. With the head and shoulders delivered, it is usually very simple to complete delivery of the lamb.

Veterinarians, of course, can do a much more professional job of removing a difficult foetus or unborn lamb. However, time is important, and due to the low value

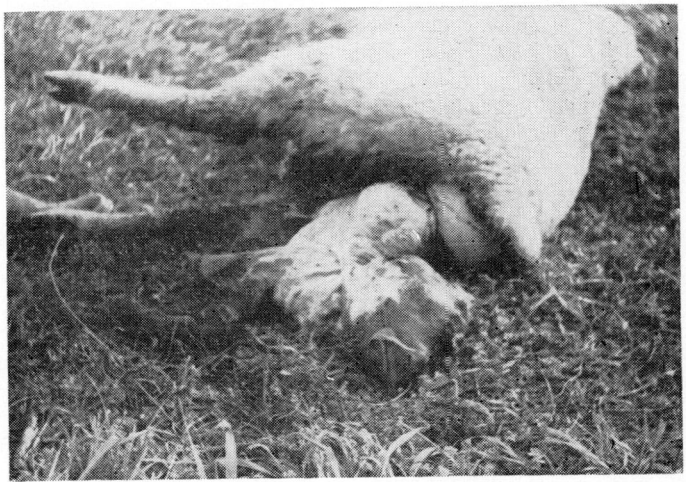

Fig. 5.6—Here is a dead ewe and dead lamb. Prompt attention could have saved both. No one was around to assist the ewe when she needed help.

Fig. 5.7—Two methods of carrying young lambs are illustrated above. The ewe will generally follow wherever her lamb is carried.

of a single lamb, it is generally not practical to call a veterinarian unless he is nearby or the sheep in question are valuable breeding stock.

5. Caring for Newborn Lambs

Make sure lamb is breathing—Now and then a lamb may appear lifeless at birth. Quick action is necessary to prevent death. As ewes may not lick the veil or mucus off the mouth or nostrils, the shepherd should remove it to prevent smothering. Then, slap the lamb on both sides of the body just back of the shoulders or rub it briskly with a burlap sack or handful of straw in order to stimulate breathing.

Disinfect navel—As soon as the lamb is breathing, his navel cord should be dipped into tincture of iodine in order to prevent navel ill or infection of the navel.

Assist weak lambs to nurse—Ewes' teats are sometimes plugged so that milk refuses to come. Therefore, squirt a bit of milk from each teat. A strong lamb has no trouble getting milk, but a weak lamb may perish if it does not get assistance at this time. Generally,

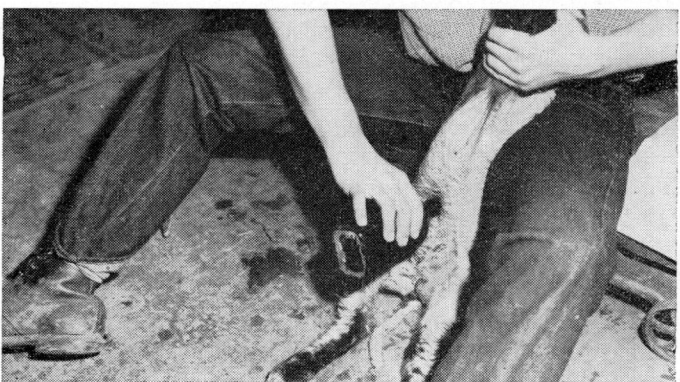

Fig. 5.8—Disinfecting the navel of a newborn lamb. Tincture of iodine is satisfactory and the entire navel and drying cord should be dipped.

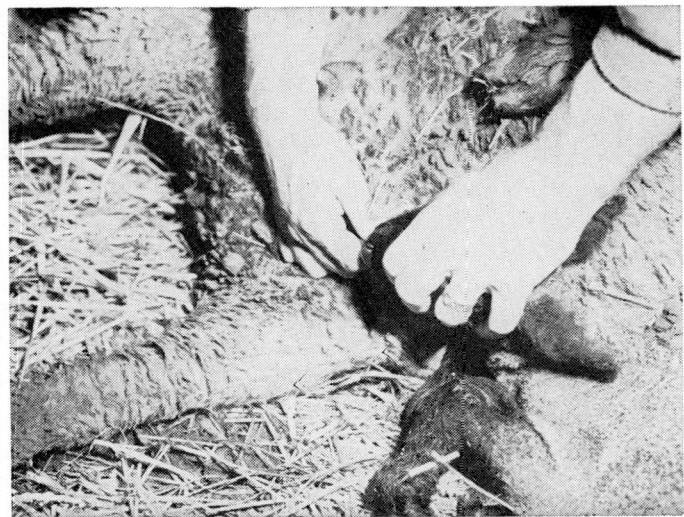

Fig. 5.9—Assisting a weak lamb to find the teat and nurse. Such a practice may have to be done once or twice until the lamb is strong enough to nurse by itself.

holding the lamb to the teat does the trick, but in extreme cases the ewe may be laid on her side so the lamb has less difficulty in finding the teat. Sometimes the first milk must even be drawn from the udder and fed with an eye dropper.

The first milk (colostrum) is very healthful for newborn lambs. It is surprising how a little milk will put life into a very weak lamb. Therefore, every effort should be made by the shepherd to see that each lamb, particularly if weak, begins nursing. Some sheepmen recommend tickling the lamb under the tail to stimulate him to suck.

Warm chilled lambs—Lamb brooders, (see Chapter VII), well-bedded jails, sheltered lambing quarters, etc., will do much to keep newborn lambs warm in adverse weather. However, when these facilities are inadequate, the lamb should be dried and wrapped in

RAISING LAMBS

sacks. Extreme cases will warrant dipping lambs in warm water until thoroughly warm. Water temperature should be such that a person can barely hold his elbow in it. **Dry well** and wrap in sacks afterwards. It is advisable to dry and rub lambs briskly in order to start circulation.

Save orphaned lambs—These lambs are frequently a problem but well worth saving. Commercially, it is seldom feasible to raise these lambs by hand; however, there are a number of practices which will save orphaned lambs by getting other ewes who have lost their lambs to adopt them.

Disguise lamb's odor—A ewe recognizes her lamb by smell, so sprinkle some of her milk on the lamb's rump and on the nose of the ewe. In some cases, tying the ewe with a small halter for a few days with the

Fig. 5.10—This arrangement illustrates how a sack and 2 x 4 can be used to prevent a ewe from turning around while orphans are being "grafted" onto her. The pole is put over the rails in a small pen in order to restrain her, although she bears her own weight on her feet. Cut both ends out of the sack and place the ewe so the teats and udder are exposed for the lamb to nurse.

Fig. 5.11—The brooder illustrated in the corner is warmed by a 40-watt bulb. Warm, dry pens like the one above give lambs a healthy start.

lamb in the same pen is all that is needed. Frequently, simple procedures like sprinkling **talcum powder** or **perfume** on the nose of the ewe and on the body of the lamb will disguise the lamb enough until the ewe learns to claim it for her own.

An orphaned lamb may be given to a ewe who has lost her lamb by skinning the dead lamb and allowing the orphan to wear the pelt for a couple of days.

Inspect frequently for sore eyes—Rolled-in eyelids should be "worked out" several times daily. According to **Michigan State Extension Folder F-19,** persistent cases may be sewed back. Wash the eyes with 15% argyrol.

6. Handling Ewes After Lambing

Keep the ewe under close observation for several days after the lamb is born.

RAISING LAMBS

Fig. 5.12—Twins are often tied together by the front leg in order to prevent their getting separated. In large flocks especially, the ewe may take one of the lambs and not bother if the second one becomes lost or is at the far side of the flock. On the range, ewes may even be tied for a brief period to a tree, brush, or whatnot until the lambs have had a chance to nurse several times, dry off, and get accepted and used to their mothers. Anything can be used to tie their legs together: string, small rope, rag, etc., but a slipknot should be put at each end so the cord can't slip off. Twelve to 15 inches apart is about the proper distance between lambs. Sometimes it may work to tie a bummer or orphan lamb to one with a mother in order to get her to accept the stranger.

Fig. 5.13—An orphan lamb makes a good pet and can direct growing children into a worthwhile, rewarding, and educational experience.

Fig. 5.14—Upper picture shows the skin of the ewe's dead lamb tied over an orphan. Notice the herdsman's hand nearby as the ewe was still uncertain. Lower picture is the same ewe and lamb 24 hours later, which demonstrates that the ewe has accepted the lamb as her own.

RAISING LAMBS

See that ewe "cleans"—Keep the ewe comfortable and warm. If her bowels and udder are in good condition, it will aid in helping her "clean" herself or expel the afterbirth. If she has no appetite, after several days she should be given a physic of four ounces (one-fourth pint) of raw linseed oil or two to four ounces of Epsom salts in a pint of warm water. A rectal injection of one quart of warm, soapy water is good in case the ewe is constipated.

Examine udder frequently—Lambs should be taking all the milk. If not, udders should be milked out. Sore

Fig. 5.15—Branding a young lamb.

teats should be treated as previously outlined in this chapter, as a ewe with sore teats may not let the lambs suck. Vaseline or commercial preparations can be used to grease teats. If the lamb's mouth also gets sores, such lip sores should be painted with iodine or creosote dip.

Brand ewe and lambs—In order to prevent mixups and find mothers easily, many sheepmen make it a practice to brand a number on the back of each lamb and the corresponding number on the ewe's rump. Branding numbers of this purpose can be made of heavy (one-fourth inch) wire or purchased through the state Wool Growers Association. The smaller the flock, the less important this practice becomes.

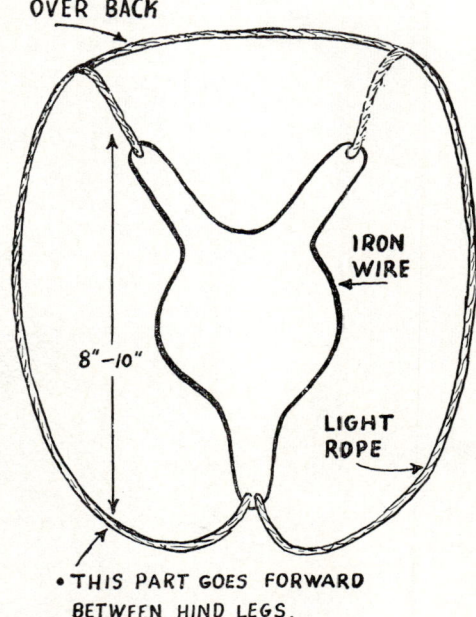

Fig. 5.16—A foxhead used for mechanically retaining a uterus. The foxhead is made of $3/16''$ or $1/4''$ diameter iron wire. Dotted lines indicate the rope which holds it in place.

Separate twins from singles—In small flocks this is unnecessary; however, in large bands it is a desirable practice to jail twins and their mothers for five or six days until they are attached to each other. When herders are rushed, they sometimes use a piece of stout cord (three feet long) and tie one front leg of each lamb to the other until they have time to make other preparations.

Observe care of everted womb—Occasionally, due to severe straining, the womb may protrude its full length (eight to ten inches). Prompt attention may save the ewe, but for the first time a beginner should observe an experienced operator replace the womb. This consists of thoroughly cleansing all parts and then pressing the parts back into place while a helper elevates the rear quarters. After this, the vulva is tied or a foxhead utilized to hold uterus in place.

Fig. 5.17—Ewes and their lambs on good pasture. Lambs eat grass at an early age, but it must be high quality.

Feed ewes light at first—After ten days ewes generally will be normal and able to take grain if the flock owner intends to feed concentrates. Most sheepmen, however, prefer to feed high quality legume hay or place the ewes on good pasture. Ewes will regain their appetites gradually.

7. Feeding Young Lambs

To bring top prices, early spring lambs must be fat and weigh 75 to 90 pounds.

Keep ewe in good flesh—There is no feed like the ewe's milk for putting rapid gains on young lambs. For this reason, it is imperative that the ewes be maintained in good shape, as the lamb will reflect directly the condition and health of its mother. Good pasture, especially legume pasture, is a must for ewes if they are to remain in good shape. Additional grain and hay will pay dividends if it is necessary to keep the ewe in good flesh. One-half pound a day is generally ample. **Farmers' Bulletin 1710** recommends that ewes be fed the following:

Ration Before Lambing			Ration After Lambing		
Alfalfa hay	4	lbs.	Alfalfa hay	5	lbs.
Whole oats	½	lb.	Corn silage	2½	lbs.
			Whole oats	½	lb.

Feed lambs fresh pasture—Lambs will begin to eat some grass at a very early age and will be eating a considerable amount by several months of age. Pastures must be better for lambs than for ewes. Therefore, many sheepmen prefer some kind of arrangement that will permit the lambs to graze ahead of the ewes. Others prefer some sort of fencing arrangement so that the sheep stay in one field not over three or four days and not return to that field before ten days or two weeks.

8. Creep Feeding

Creep feeding is probably unnecessary for most producers. The greatest percentage of lambs that go to market are not creep fed. However, for show lambs or whenever unusually poor range or pasture conditions prevail, it is advantageous to creep feed lambs because they must be kept fat in order to bring high prices. The ewes, on the other hand, can go down in flesh, but will be unharmed when the yearly cycle is considered, as they will regain any loss when feed conditions improve.

Creep feeding seems to be most significant when there is a question of whether or not lambs will get to market early enough to command higher prices than those marketed later.

Ohio State University reports some interesting results in this regard as follows:

"The inclusion of soybean oil meal in the creep ration once again resulted in faster, more economical gains. There would seem to be little question as to the profitableness of creep feeding compared to no creep, in this year's test, when measured by the additional gains and

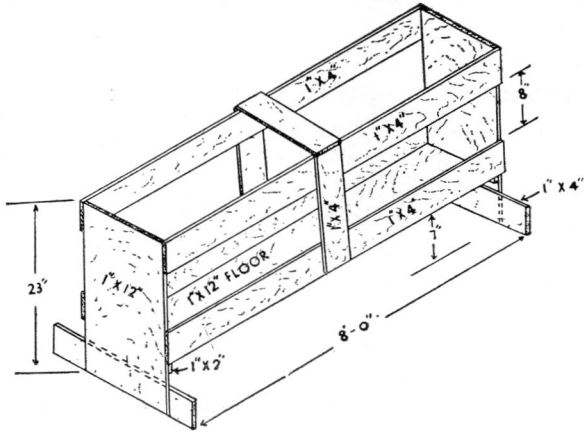

Fig. 5.18—A grain bunk of this style is used in the lamb creep.

Fig. 5.19—Creep fed lambs like these not only gained more per day but brought more per pound when sold because they went to market sooner.

higher prices received for early marketed lambs."

The lambs marketed in June brought $2.00 less per cwt. than those marketed in May; and in July, when the first of the non-creep lambs were marketed, they brought $1.50 less per cwt. than those marketed in June.

Ohio State University further reported:

"At weaning time the non-creep lambs were gaining less than one-tenth pound per day while the creep fed lambs were gaining at the rate of from one-half to two-thirds pound per day. It is anticipated that the feed required, after weaning, to finish the non-creep fed lambs will be in excess of all the feed required to finish the creep fed lambs. This would indicate further the profitableness of creep feeding under conditions similar to those of this test.

"As anticipated from last year's results, the addition of corn and soybean pellets, to the alfalfa pellets, about two months after the lambs went on creep, resulted in increased and more efficient gains."

When the individual operator decides his feed conditions merit, then it is a desirable practice to creep feed his lambs. Lambs will eat some grain at two weeks of age. Equal parts of cracked corn or rolled barley and crushed oats make a good grain mix. Bran is a good addition up to one-third of the mixture. Linseed oil meal or soybean oil meal may be added at the rate of one pound to ten pounds of grain. Place only a little grain in the bottom of the trough at first. After the lambs are eating well, the oats may be eliminated. If the lambs are on drylot, they should have access to good legume hay as well.

Many types of creeps are feasible. (See Chapter VII.) Creeps made with rollers will permit rather large lambs to enter and yet exclude the ewes. Smooth lumber should be used at all times in order to protect the wool.

9. Preventing Diseases of Lambs

The chapter on diseases and parasites also includes most maladies that will affect lambs. However, there are a number of troubles that particularly bother lambs, but adequate precautions can materially reduce these dangers.

Warm chilled lambs—During cold weather, chilled lambs will die if not dried and warmed. A good procedure is to wrap them in woolens after they have been wiped dry, and place them in a basket. It may be necessary to place warm bricks or a hot water bottle in the basket with them. Severe cases may be warmed by completely immersing the lamb in a pail of water warm enough for the arm to stand comfortably. Hold the neck and head out, of course, and dry thoroughly after the lamb is warm. Do not keep lambs away from their mothers any longer than necessary.

Remove dung from tail—Feces that stick to the lamb are referred to as "pinning." If the condition becomes acute, it will prevent the lamb from voiding. Therefore, all fecal material should be removed, and then a handful of dust or sand should be thrown on these parts to prevent a repetition of this condition.

Watch for scours—If scouring is due to overfeeding, the lamb will lose as much weight in two days as it will gain back in two weeks. Therefore, scouring is serious and producers must be continually on guard. The first practice, should scouring appear, is to cut down on concentrates although it is not advisable to completely eliminate concentrates if lambs are accustomed to heavy feeding. The ration should be cut at least in half. Young lambs may scour if the ewe is off feed, if a sudden change is made in the ewe's feed, or if the ewe gives too much milk. Soiled teats due to dirty pens may be another cause. Strict hygiene, plenty of clean, dry bedding, clean, dry pastures, rest and quiet are important in preventing and treating scours. Removing any of the causes generally cures the scours unless the infectious stage has become serious; then the services of a good veterinarian should be obtained, as the condition can spread to the remainder of the flock.

The feeding of antibiotics, such as penicillin, while giving excellent results with pigs, calves or broilers, has proven thus far to be of little value even for weak, sickly, or scouring lambs.

"Work out" rolled-in eyelids—Several times daily rolled-in eyelids should be worked out if the lid is not too severely affected. Persistent cases must be sewn back by taking a stitch in the eyelid to cause it to remain in the correct position for a day or two. Another method is to nick the eyelid with a sharp knife. (Do not cut entirely though.) When the wound heals, scar tissue will pull the lid out of its curl and remove the inflamation caused by eyelashes getting into the eye. Wash the affected eyes with a 15% argyrol solution.

Treat sore mouth—There are many kinds of mouth ailments common to young lambs. Some of these are quite serious, as death from starvation may be a result. Their prevalence may vary considerably in different parts of the country. If usual methods do not stop and prevent spread of sore mouth, the services of a veterinarian or a competent animal husbandman should be obtained at once.

Cornell Extension Bulletin 828 says that many cases of sore mouth may be cured by applications of 10% nitric acid solution, tincture of iodine, or a saturated solution of potassium permanganate to the affected parts. The scabs should be removed and sores opened before the treatment is given. Some types may be prevented by vaccination, and this practice should be employed if repeated outbreaks occur.

Dip navels—Navel ill is a disease caused by infections which enter the body through the navel cord. Lameness or a swelling of one or more joints of the legs, such as knees or hocks, is characteristic. The best prevention is to dip the navel cord in iodine solution soon after birth, as once the animal contracts the disease, it is a case for a veterinarian. However, commercial lambs are seldom worth this expense unless large numbers are involved. Joint stiffness or navel ill is not to be confused with the following disease.

Prevent stiff lamb disease—Much has been written about this disease, and a great deal of study has been carried on by various experiment stations. Most stations agree that common stiff lamb condition is a result of the ewe's ration. Cornell reports that stiff lambs can be produced by feeding a ration of cull beans and alfalfa hay. No stiff lambs result when ewes are fed a ration of oats, wheat bran, corn silage, and mixed hay. There-

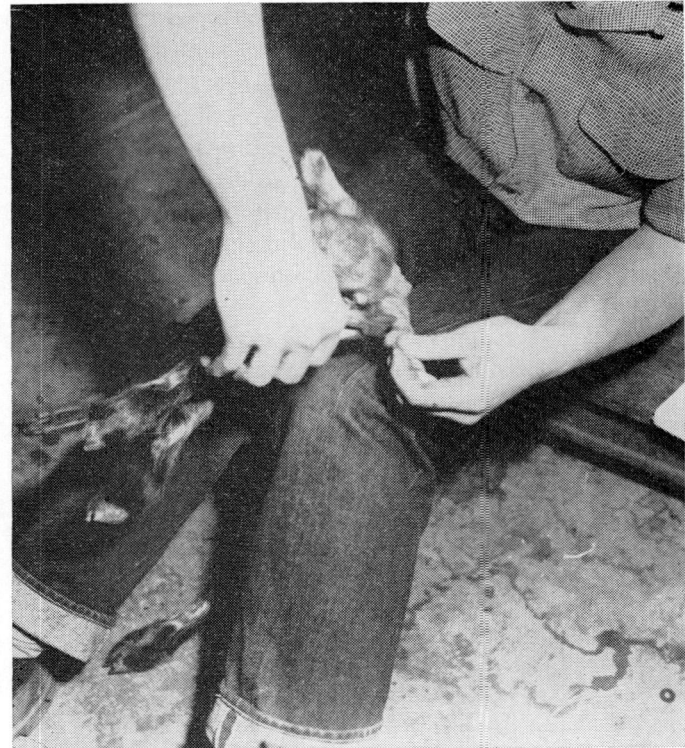

Fig. 5.20—Ear notching sheep is one effective way of identifying lambs. This method eliminates the use of metal tags. (See Fig. 3.7 and Fig. 5.21.)

fore, it is desirable to include grains in the ewe's ration that are high in vitamin E and carbohydrates.

Young suckling lambs two to six weeks of age are most often affected. They have difficulty in walking, and seem paralyzed in both front and rear legs. Some lambs recover. A veterinarian should be called if the condition is at all prevalent, and his recommendations followed in treating young lambs.

Curb constipation—One teaspoonful of castor oil for constipation will generally relieve the trouble.

RAISING LAMBS 167

10. Weaning Lambs

Weaning lambs is seldom a problem. However, it is important that they do not lose the milk fat they have obtained from their mothers. Hothouse lambs and early spring lambs may never be weaned from their mothers but go directly to market off the ewe. Many lambs will wean themselves by eating more and more pasture as the ewe dries up.

Wean at 3½ to 5 months of age—Most lambs should be weaned at four months of age. There is no need to wean lambs too early if the ewes are milking well and in good condition. On the other hand, if the ewe is giving little milk, the lamb will do better if taken off and put on better pasture. The ewe also will maintain better health.

Separate ewes from lambs—Do not allow them to run together, as a complete break or separation is the easiest and most effective practice when weaning.

Give lambs best feed—Weaning is a mild shock so

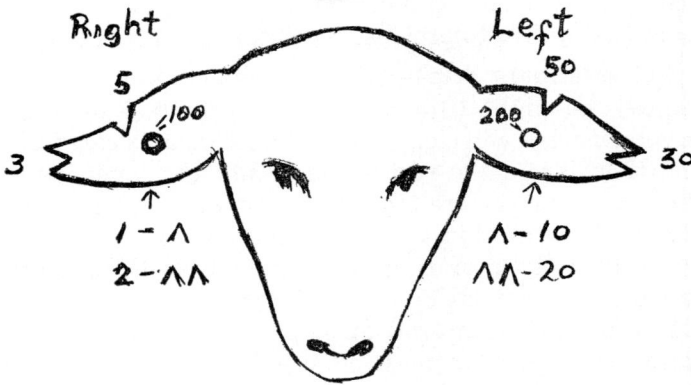

Fig. 5.21—One system of ear notching sheep. This system is used on the flock at the University of California, Davis, California. One or two (10 or 20) is formed by either a single notch in the lower side of ear for one, or two notches side by side for two. Holes are punched to indicate hundreds.

that everything possible should be done to cushion its effect. New pasture is excellent and should it not be available, supplemental feeding with grain may be advisable for a time. Be sure ample water and salt are always on hand.

Check ewe's udder—Generally, no trouble ensues. However, most sheep raisers consider it a desirable practice to check the udders every other day. Severe cases of caked udder should be milked out. Chalk mark these cases and watch for several days until they are dry. Do not milk out slightly swollen udders as this condition is normal.

11. Preventing Prolapse of Rectum

In this situation the rectum of either male or female lambs may protrude out the rear from two or three inches to eight or ten inches in severe cases. Fortunately it does not happen frequently although in some years and with some flocks it may be more prevalent than with others. Lambs strain, get weak, seldom eat and will die if not attended to. Ewe lambs seem to be more susceptible to prolapse of rectum than ram lambs.

Get veterinary help—After a sheepman has seen the remedy, he might attempt to perform the task himself. A veterinarian will clean the protruding parts carefully and then insert the rectum back into the lamb while the rear quarters are slightly elevated. A strong cord is sewn into the skin around the rectum opening in the shape of a circle or ring small enough so the rectum cannot be forced out by the lamb straining, yet leaving a hole for fecal material to be excreted. Frequently a cauterizing solution may be injected into the tissue surrounding the rectum so the walls of the rectum will thicken. As the lamb heals, the size (diameter) of the ring may have to be altered and eventually the cord removed.

Dock long—Some persons well acquainted with this condition suspect that docking sheep too close, as is often done with ewe lambs, may weaken the muscle attachments holding the rectum in place and thus allow it to prolapse. Therefore, it is a desirable practice to dock ewe lambs as long as ram lambs in an attempt to prevent prolapse of the rectum.

12. Controlling Predators

In many areas predators are one of the major sources of loss in raising lambs. This is caused in part by their tendency to bunch together when attacked and their complete helplessness from preying enemies. To some degree cattle or other large livestock try to fight off predators especially when they have their young, but sheep cannot even defend themselves. On the range or in sparsely settled areas coyotes, bobcats, mountain lions, occasionally bears, or even eagles prey on sheep, especially lambs. However, coyotes are most numerous and cause by far the most damage.

In more populated areas, particularly with farm flocks, dogs cause a great deal of damage. Unfortu-

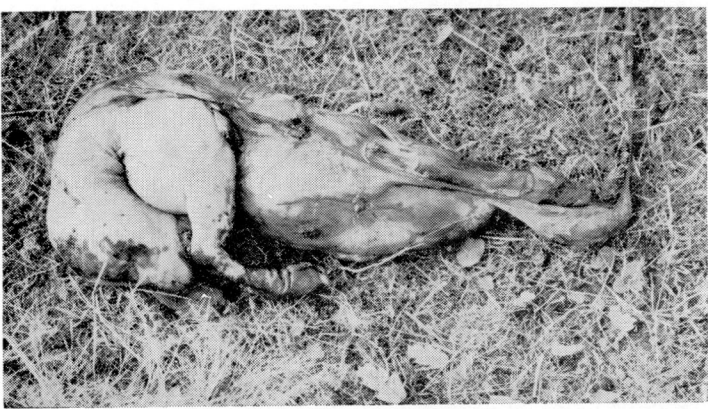

Fig. 5.22—An aborted lamb. This lamb was aborted because the ewe was harried by dogs just before she was ready to lamb.

nately, a household pet may be the culprit responsible for damage in as many cases as a pack of semi-wild unowned dogs.

Naturally, such a situation can cause a great deal of hard feelings or sorrow as owners of the dogs refuse to believe the damage done by their dogs even when confronted with proper evidence. Furthermore, a small, friendly pet under the right conditions is just as likely to kill sheep as a large, vicious watchdog.

Dogs and coyotes constitute a real problem in many areas, especially for farm flocks. Coyotes seem to thrive near civilization so oftentimes both may be preying on the same flock.

Examine the evidence—Coyotes and other wild animals kill for food so generally the sheepman will lose a single sheep per night or "kill." The sheep, or more often a lamb, will be eaten beginning with the entrails or heart and other organs. Teeth marks will be evident around the face and throat. Many times the wool and skin will not be torn up as is the case with dog injury. Killing is a vice with dogs so usually a considerable number of sheep will be torn up, injured or maimed even though only a few or none may be actually dead. Damage is so serious that many eventually die or have to be killed.

Dogs do not kill for food so the carcasses are rarely, if ever, eaten. Few teeth marks are found on the face and throat as most injury consists of large patches of skin and wool torn from the sheep with lacerations and deep cuts on the legs and flanks.

In all likelihood dogs start chasing sheep as innocent fun. The sheep run and dogs nip at them resulting in the teeth getting caught in the wool and tearing the skin. Eventually they get a taste for blood and become sheep killers, killing for no apparent reason and going out of their way to find a flock. Seldom, if ever, can a

Fig. 5.23—A lamb that has been killed by dogs. Note the patches of skin and wool torn from the sides and flanks with little injury to the head. In addition, no portion of the dead animal has been eaten. This is almost certain evidence the damage was caused by dogs. Coyotes or other predators always eat part or most of the carcass, generally starting with the entrails.

dog be broken of this habit once it is acquired. Any dog seems susceptible, even those raised with a flock, and many working sheep dogs are tied when they are not actually working sheep with the owner.

Check the law regarding dogs—Many states and counties have very specific regulations concerning dogs left to run at large. In addition, liability and compensation are often spelled out so both the owner of the dog and the sheepman know how they stand in the event of loss. It is important to know when and how dogs can be exterminated so that sheepmen do not become enmeshed in a countersuit.

Notify authorities—Should damage occur, the sheriff or other appropriate authorities should be contacted immediately and asked to examine and inspect the damage. Their suggestions may be helpful and oftentimes their written report can be the basis for getting reimbursed for sheep killed by dogs. Take pictures, as photographs help establish a claim, because sympathies from the general public are often on the side of the dog owner should a case go to court.

Check your sheep association—Sheep or wool associations can supply excellent advice on how to cope with specific cases. Many times the association may suggest legal advice or recommend where it might best be obtained locally. The association often has a record of other breeders who have had similar difficulties. These breeders can be contacted to see how they went about stamping out the problem.

Try joint action—It is expensive for a single sheepman to fight the predator problem alone. If the damage is caused by dogs, legal advice and other expenses can be shared. Many sheepmen working together can better pinpoint the source of trouble. If the predators are coyotes, state trappers or other agencies will respond to a group more readily than to an individual. Occasionally, it may pay to hire a specialist to trap an especially wily animal.

Determine which predator is responsible—Sheepmen may find that a single coyote or dog is responsible for almost all damage at any one time even though many other dogs or coyotes or other predators are in the vicinity. Most sheepmen would say the predator, regardless of what kind or to whom it belongs, should be killed on the spot. County or state laws should be followed rigidly in this regard.

Use coyote guns—Even though a sheepman may wait many days or nights, the one time he relaxes his vigil may be the time the marauder strikes. For this reason coyote guns properly used and with adequate safeguards are excellent coyote killers. The gun shoots a

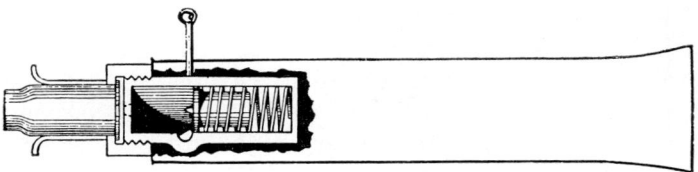

Fig. 5.24—Cross section of humane coyote getter. This is an effective device for killing coyotes. The instrument is buried in the ground to about the level of the trigger. A bit of rabbit fur or wool is wrapped around the upright cartridge and then baited with a few drops of scent. A chemical kills the predator. (Courtesy, Humane Coyote Getter, Pueblo, Colorado)

cyanide pellet into the coyote's mouth. Use the gun according to manufacturer's directions. Even though the gun is relatively safe from grazing stock, it should be kept away from areas where livestock or people could easily touch it.

Make your own bait—Prepared baits are available, but often one is not on hand when it is wanted. A simple bait to use with the cyanide gun may be made by filling a pint jar half full of bits of liver, spleen or other organs, hamburger, cheese, fish or entrails. Add about 20% water, and screw the lid on. Let the bait set for about one week. An evil smelling liquid results, but a small amount placed on the trigger is effective in drawing coyotes to the gun. Sets should be rebaited every two or three days.

Enlist aid of sportsmen—Wild animal game calls are very effective in the hands of experienced sportsmen in calling coyotes and other predators to them. Frequently, local hunters will gladly come to a farm or ranch and call for wild animals as a form of recreation, and thereby assist the rancher in getting rid of sheep killers.

Pen at night—Many ranchers do this as a general rule even when there is little danger from predators. During lambing season it is advisable to have a field close to farm headquarters so ewes and lambs can be watched constantly. Under these conditions sheep are protected and predators, especially coyotes, seldom give trouble. Therefore, if losses persist, it may become necessary to "pen" sheep every night until the source of trouble can be eliminated. A little grain fed each night will aid in getting sheep to come in, thus making this task a simple one. Electric lights installed by the pens will give additional protection.

CHAPTER VI

FEEDING AND FATTENING SHEEP

As with other types of livestock, with sheep the greatest single expense item for yearly running is their feed. For this reason, careful planning and use of feedstuffs is imperative if one is to reap the most profits from the sheep business. Sheep feeding does not receive enough attention by many because sheep tend to eat a high percentage of weeds and other coarse feeds.

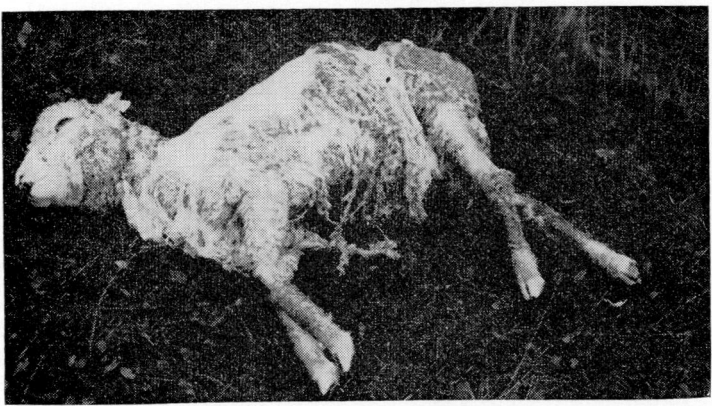

Fig. 6.1—This ewe died from many causes; however, the primary cause was malnutrition. When she became thin and poor, a "wool break" occurred so the wool was shed allowing other diseases to destroy her. Proper feeding will insure healthy productive sheep.

It is estimated that approximately 95% of the feed of sheep is obtained from roughages. Even with many farm flocks, grain is fed only during lambing season. While sheep eat many more coarse feeds, they will always do better on good pastures and properly balanced rations. Proper use of pastures and by-products as well as sound, economical buying involve many approved practices that sheepmen should apply.

Activities Which Involve Approved Practices

1. Supplying desirable pastures.
2. Selecting desirable concentrates.
3. Providing adequate minerals and vitamins.
4. Providing sufficient water.
5. Providing winter rations.
6. Using antibiotics.
7. Obtaining good buys.
8. "Sheeping down" for fattening.
9. Utilizing by-products.
10. Feeding out by contract.
11. Feeding for show.
12. Creep feeding lambs.
13. Pelleting feeds.

1. Supplying Desirable Pastures

Sheep are better adapted to utilize a wider variety of plants than other kinds of farm livestock. They also do extremely well on irrigated pastures and are marketed as fat animals off such fields. Good pasture management will provide a maximum amount of feed capable of producing maximum growth.

Feed your pastures—The first step in producing the right kind of pasture is to be certain that the soil on which pasture is growing contains the proper plant foods. Pastures can be no better than the soil on which

Fig. 6.2—Sheep utilize a wide variety of plants. This band of several thousand sheep is migrating from permanent pastures in the lowlands to high mountain range.

they grow. Therefore, proper nutrition and a balanced diet for sheep grazing on pastures depend to a large extent on the soil fertility. In addition, yield of pasture and control of erosion is enhanced by properly fertilizing pastures. **Illinois Circular 649** states that the value of the legume-grass crop in improving soil and controlling erosion depends on the amount of top and root growth. This, in turn, depends mainly on the amount of lime, phosphate and potash in the soil. No two soils are alike in what they lack. Some newly-farmed soils may lack nothing. The Extension Service or agricultural colleges can assist materially in any local situation.

Fertilizing pastures is not confined to irrigated pastures; as rangelands, especially when reseeded and fertilized, show remarkable results. Dr. Merton Love of the University of California at Davis reports excellent results from reseeding rangelands with desirable legumes and fertilizing with 200 pounds or more per acre of superphosphate. Adequate rainfall must be present.

Some results indicate that livestock prefer to eat grasses and legumes grown under unfertilized soil con-

178 APPROVED PRACTICES IN SHEEP PRODUCTION

Fig. 6.3—An excellent stand of legume and grass pasture. The pasture mixture is ladino clover, Kentucky 31 Fescue, and lespedeza. This feeds 120 head of sheep on 10 acres. The photo was taken in Kentucky—Class I and II lands, Group 20 soils. (Courtesy, Soil Conservation Service)

ditions. This may be because of too rank growth produced by fertilizing. However, with the amounts of fertilizer commonly recommended, this factor is not present and the tremendously increased tonnage produced makes fertilizing pastures a very desirable practice.

Get the right mixture for your locality—No one pasture will do for every locality. This applies to reseeding rangelands as well as irrigated pastures. Alfalfa brome is a recommended pasture in many parts of the midwest. Ladino clover mixed with one or more of the grasses such as orchard, rye, or tall fescue is common in the irrigated areas west of the Rockies. Straight ladino is also used, but increases the danger of bloat. Different varieties of trefoil will do well under heat or soil conditions not too favorable for ladino. Alsike and strawberry are common in northern states. The best

FEEDING AND FATTENING SHEEP

procedure is to consult your county agent or agriculture college and follow their recommendations on your farm. Generally it is advisable to try the mixture first on a small acreage of your farm and abide by these results.

Provide year-round grazing—Pasture is most often the cheapest source of nutrients for sheep. If the grazing season can be extended, more economical gains and reduced labor costs can be obtained. The use of new varieties, fertilization, etc., of pastures are possible ways to extend the grazing season. Development of springs, proper salting, and fencing also make it possible to use a higher percentage of the available feed on the range. Many sheepmen leave certain fields ungrazed until fall or winter or else graze them sparingly in the spring so winter pasture is available.

Supplement on poor pastures—During poor feed years or on inadequate pastures, grain or protein supplements will pay big dividends. Generally, only a fourth to one-half pound of protein supplement or a pound or two of grain per day is all that is necessary. Feeding at this time not only maintains weight and

Fig. 6.4—A top pasture like this would not have to be supplemented, although bloat is a danger. Many pastures need to be supplemented with additional feed for sheep.

Fig. 6.5—Trees provide excellent shade for sheep.

health of the sheep, but reduces the strain on the pasture.

Graze cattle and sheep together—This is not a must as many ranches are set up to handle only one or the other kind of livestock. However, this recommendation is standard practice in other countries, notably New Zealand, where it has proved desirable and profitable. The idea is that cattle "wrap their tongues" around tall grasses; the sheep clip the short species. Thus all forage is utilized equally. Pastures must not be overgrazed but properly rotated.

Stock fields properly—There is no way of arriving at a single figure for all fields that would be the correct carrying capacity. Some fields are capable of carrying five to ten head per acre; others would take as much for one sheep. Most good producers like to keep the feed a little better than the sheep, that is, understock rather than overstock. This is especially true if lambs are being fattened on pasture. On the other hand, ewes can get along quite well on relatively poor pastures if they are not nursing lambs. In western regions where Mediterranean type plants prevail, overstocking on annual legume and grassland pastures is desirable once these plants have set seed.

Provide shade—In arid, hot regions, grazing sheep need some place to get away from the sun. Under these conditions, they feed mostly in morning or late afternoon and will lie down during the hottest part of the day.

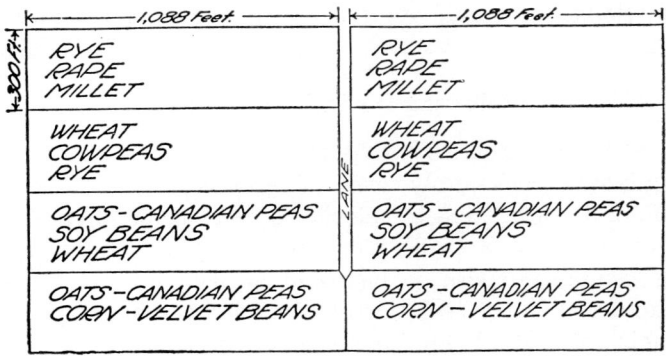

Fig. 6.6—Diagram of a farm showing a practical rotation of crops for each field in Central Atlantic and Corn Belt states.

Be alert when fattening on pasture—Thousands of lambs and wethers are fattened on irrigated pastures throughout the country each year. Oftentimes the margin of profit is slim. Lambs lost from bloat can be a major factor, as the pastures used always contain a high percentage of lush-growing legumes. Someone must be on guard constantly to watch for danger signs and treat animals in need of help.

Use temporary pastures—Any annual crops that are sown with the idea of being pastured off before maturity are called temporary pastures. **Farmers' Bulletin 1181** has some excellent information on temporary pastures and recommends them for the following reasons:

1. They make it possible to fit the flock into the livestock farming system with little change in the usual method of producing feed for cattle and swine.
2. A uniform flow of milk with ewes is maintained.
3. Internal parasites are greatly reduced.

Some crops that make excellent temporary pasture are fall sown wheat, rye, barley, and alfalfa; spring sown oats, peas, rape, cowpeas, Italian ryegrass, soybeans, and corn. Sudan grass is also a top-notch feed, particularly in arid regions. Surplus feed from any of these crops makes excellent hay for winter roughage.

2. Selecting Desirable Concentrates

Sheep are able to utilize most grains and other concentrates reasonably well. Therefore, the important factor is to be able to select the most economical feed. However, some feeds work better than others for a particular purpose so it would be desirable to understand the characteristics of concentrates in relation to their value as sheep feeds. **Texas Bulletin 129** has this to say about the following concentrate feeds for sheep feeding:

Grain sorghums and corn—These have about the same feeding value, although corn is preferred, and are the chief lamb fattening grains. Threshed or shelled sorghum grains make, on the average, larger daily gains than heads either ground or whole. Milo ranks highest among the grain sorghum heads followed by feterita and kaffir.

Wheat—This is the best of the small grains and is practically equal to corn. It should not be ground for lambs when hand fed.

FEEDING AND FATTENING SHEEP

Barley—This is better used by mixing with corn, wheat, or threshed grain sorghums in proportions of three to five, or one-half to one-half.

Oats—This may be fed whole, but not as the sole fattening grain, for oats fail to produce a good finish. They are valuable in starting lambs on feed because of their bulk, palatability and conditioning value.

Grain sorghum gluten feed—This may be fed with alfalfa hay and grain. A mixture of seven parts grain and three parts grain sorghum gluten feed has given good results.

Protein supplements—Cottonseed meal or cottonseed cake (sheep size) is usually more available and recommended as a protein supplement. Other oilseed meals or cakes as linseed, soybean, and peanut may be used just as cottonseed meal or cake.

Fig. 6.7—Champion pen of fat lambs. These show lambs have been fed concentrates in addition to mother's milk and pasture.

Cottonseed—This may be fed with alfalfa and grain as a replacement for part of the grain, and also to supply additional needs for protein. Lambs will usually not eat more than about .6 pound of cottonseed daily per head.

Alfalfa—This is the most desirable hay for lamb feeding and should furnish at least one-fourth of the roughage in all lamb fattening rations.

Blackstrap molasses—While lambs have been fed up to 20% of the total ration in molasses, about 12½% in the ration will serve the purpose of laying the dust and binding small particles of feed together. It has about 70% of the feed value of corn

Silage—High quality silage can be used as the sole source of roughages in fattening lambs. A small amount of high quality hay improves the ration in most instances. When feeding grain sorghum silage, be careful to closely estimate the grain content.

Do not grind grains—There is no benefit from crushing or grinding most grains, except for old sheep with poor teeth, for young lambs up to five to eight weeks of age, or for show lambs. Occasionally ground feed is given fattening lambs being self-fed on mixtures of grain and chopped hay. **Kansas State University Circular 269** says that "Some hard-seeded grains such as hulless barley or grains with a lot of weed seed should be ground or cracked."

Supplement feed according to pasture and type of sheep—Fattening lambs vs. range sheep will, of course, have different food requirements. Sheep grazing ample green feed, or eating at least two pounds of legume hay daily, usually get enough protein; therefore, when pasture is short, it must be supplemented.

Farm sheep—Every means possible should be utilized to improve and get the most from pasture. However,

FEEDING AND FATTENING SHEEP 185

when grazing is poor, the breeding flock should be fed roughage and one-fourth to one-third pounds of cottonseed meal or cake daily. Add one-half pound of grain to ewes' rations a month or two before lambing. The meal should be increased to one-half pound daily after lambing. If good legume hay is fed, only grain is needed.

Range sheep—While range sheep get most of their nutrients from well-managed pasture, when grass is short or dry they should be fed cottonseed cake in order to supply their bodies with necessary protein and phosphorus. A desirable practice is to start feeding early so sheep do not go downhill before going into winter. Start in the fall with one-tenth pound of cake daily per head, gradually increasing the amount as winter progresses. Up to one-half pound may be fed depending upon the severity of the winter. Supplemental feed should be reduced gradually as spring grazing becomes available.

Fig. 6.8—Range sheep can get most of their nutrients from good pasture.

Fig. 6.9—Fat lambs at market. Lambs should be sold as soon as they are fat.

Lamb fattening—Many sheepmen may never fatten a lamb other than those fattened by mother's milk. To others, fattening lambs with concentrate feeds is their main enterprise. Most feeders who show a profit feed lambs each year and carefully utilize all feeding principles.

Feed light at first—For whole grains and pea-sized cottonseed cake, a table is provided which may be used as a guide. It shows pounds of grain and cottonseed cake to feed per 100 lambs, after a rest period of one to seven days. Roughage is to be full fed after each feed of grain and cake.

At the end of two weeks, lambs will be eating one-third pound daily of cottonseed which is full feed. Good hay should be fed free choice, but not wasted. Grain can be gradually increased until some may be eating as much as two pounds per head daily.

FEEDING AND FATTENING SHEEP

INITIAL FEEDING GUIDE PER 100 LAMBS

Days		Grain (shelled or threshed)	Cottonseed Cake (43 per cent protein)
1st	PM	5 lbs.	5 lbs.
2nd & 3rd	AM	5 lbs.	5 lbs.
	PM	5 lbs.	10 lbs.
4th & 5th	AM	10 lbs.	10 lbs.
	PM	10 lbs.	10 lbs.
6th & 7th	AM	10 lbs.	10 lbs.
	PM	10 lbs.	15 lbs.
8th & 9th	AM	10 lbs.	15 lbs.
	PM	15 lbs.	15 lbs.
10th & 11th	AM	15 lbs.	15 lbs.
	PM	15 lbs.	17 lbs.
12th & 13th	AM	15 lbs.	17 lbs.
	PM	15 lbs.	17 lbs.

Get lambs to market weights quickly—Lambs should be sold as soon as they are fat. Top lambs may go to market as much as 15 days before the main group. (A lamb is said to be fat when it is difficult to feel the backbone and rib with the fingers.) As a rule, 90 to 105 days are average feeding periods. Good lambs will gain up to one-third pound per head per day under good management.

3. Providing Adequate Minerals and Vitamins

Sheep and cattle are very similar in their mineral and vitamin requirements. Vitamin A is probably the only vitamin that would be lacking in normal feeding of sheep, as their outdoor life provides ample Vitamin D. While both minerals and vitamins are used only in small amounts, their inclusion in the diet is very

RATIONS*
(Suggested Daily Rations)

Dry Ewes	October 1 to Lambing	Lactation Period	Flushing Period
1. Legume hay 3½-4½ lbs. 2. Legume hay 2 lbs. Silage 3-4 lbs. 3. Legume hay 2 lbs. Ground dry fodder 2-3 lbs. 4. Legume hay 2 lbs. Grass hay 2-3 lbs. 5. Silage 3-4 lbs. Dry roughage (Non-legume) 2 lbs. ⅛-¼ lb. Protein supplement 6. Good pasture (Native or planted) 7. Cereal crop pasture	1. Use any of the rations listed for dry ewes. Add ½ to 1 pound of grain daily, depending upon the size, age and condition of ewes; also on the quality of roughage used. Use whole grain if ewes have good teeth. Add a protein supplement if legume roughage is limited or low in quality.	1. Use any of the rations listed for dry ewes but add 1 to 1½ pounds grain per head per day to produce milk.	1. Turn ewes to lush pasture. Use ½ to 1 pound of grain for 2 weeks before the breeding season. Grain sorghum, barley, corn or oats may be used.

Ewes require 4½ to 5 pounds dry feed or its equivalent per day.
Mineral mix of equal parts of limestone, bone meal and salt may be fed free choice.
Loose salt should be fed separately from the mineral mix. Do not feed moldy feeds to sheep.

* **Kansas State University Bulletin No. 269**

FEEDING AND FATTENING SHEEP

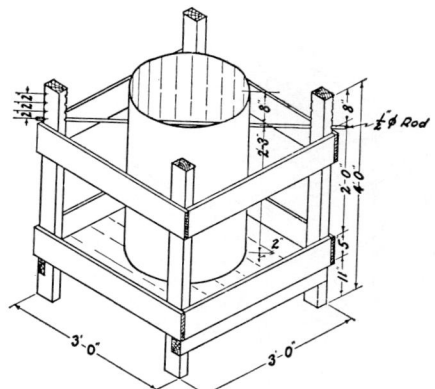

Fig. 6.10—This feeder is best used inside a barn unless it is covered.

important. Therefore, it is wise to be alert to any lack of either substance.

Watch for deficiency clues—When a piece of farm machine begins to squeak, it is already past the proper time to lubricate. The same condition exists with animals if a disease condition is indicated by some unusual hunger sign. As a general rule, sheep fed a normal ration will not be deficient in vitamins or minerals; however, sheepmen should be on guard and whenever production efficiency is below standard, a lack of vitamins or minerals should be suspected.

A low percentage lamb crop could be due to lack of Vitamin A or E. Good, green legume hay would relieve this trouble. Unusual appetites such as the sheep chewing wood, showing a preference for unusual or even poisonous plants, might be caused by a mineral deficiency. A change to a well-balanced diet will often eliminate this trouble.

Always supply salt—True, there are some of the alkali areas of the west where it is unnecessary to supply salt; on the other hand, this is one mineral that

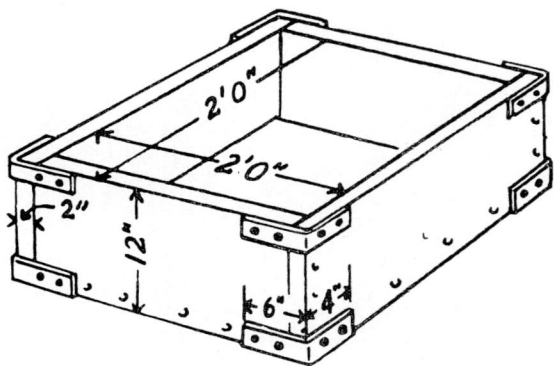

Fig. 6.11—This heavy, portable box is especially designed for the range country where the herds are moved from one feeding ground to another. It is heavily constructed of two-inch lumber and braced with metal straps. One box should be used for salt and one for a free choice mineral supplement.

should always be supplied **free choice** to sheep. Granulated salt is best, especially when there is lots of green feed available. Animals need greater amounts of salt on irrigated pastures or on green grass in the spring. Breeding ewes will need one-half ounce per head daily, and fattening lambs about one-fourth ounce per day. Block salt is satisfactory on dry range.

Feed minerals according to area—Some areas may be deficient in one or two minerals. Iodine, for example, is lacking in many areas; cobalt in others. The best practice is to consult the county agent in your particular area and add minerals to the concentrate or salt mixture according to his directions. There is no point in feeding minerals to sheep unless the minerals are known to be specifically lacking in the soil or water. Otherwise, it is a needless expense.

Keep sheep accustomed to salt—If sheep are suddenly fed all the salt they need after having been denied it for some time, death losses may occur. There is no

FEEDING AND FATTENING SHEEP

danger when salt is kept before them at all times, and it is a desirable practice.

Provide other minerals when necessary—Some feeds may be notoriously low in the two main minerals needed by sheep—calcium and phosphorus. Beet pulp is low in phosphorus and it should be added to the ration when this by-product is fed in large quantities. Non-legume forages are often low in calcium, and this mineral will have to be added unless a legume can be fed. Bonemeal and limestone are good sources of these two important minerals. Free choice is the easiest and most desirable way to supply them.

4. Providing Sufficient Water

Sheep require considerable water. It is true, however, that in the spring when sheep are grazing green grass

Fig. 6.12—Beet pulp dump near a large feeding plant. The solid mass in the foreground of the picture is 10'-15' deep. Large machinery is necessary in order to handle the vast amount of this by-product.

or whenever they are consuming high-moisture content feed, they may go weeks without water, as they often do in high mountain meadows or on spring desert grazing. Climate also affects the quantity of water consumed. Mature animals may need up to one gallon per day and lambs one-half of this amount.

Water should be as clean as possible and free of parasites. Developing springs and reservoirs will pay dividends in healthier sheep. In the winter water may have to be warmed. Occasional cleaning and chemical sterilizing of drinking equipment may be necessary. (See Chapter VIII.)

5. Providing Winter Rations

According to **Farmers' Bulletin** 840, winter management has a very important relation to the returns from the flock. Animals should be kept in good flesh and health so that wool continues to grow, even though ewes may not gain in weight.

Feed economical feeds—Leguminous hays, cornstalks, straw, or other waste roughages can form the basis of a good winter ration. Timothy hay is not a good sheep feed unless cut when immature or in the early bloom stage. One-fourth to one-half pound of cottonseed meal may be needed if all the hay is of a non-nitrogenous nature. Beet pulp, roots, or silage is an excellent winter feed and the inclusion of this into the ration will improve most diets.

Any of the following rations will provide sufficient nutrients for mature ewes:

1. 3 lbs. alfalfa or soybean hay
   2 lbs. corn silage
   ½ lb. shelled corn
2. 3½ lbs. alfalfa or clover hay
   2 lbs. corn silage
3. 3 lbs. alfalfa
   2 lbs. corn stover
4. 2 lbs. oat straw
   2 lbs. corn silage
   ¾ lb. shelled corn
   ¼ lb. linseed meal

CREEP RATIONS IN USE UNDER VARIED CONDITIONS[1]

Ingredient	kg[2]

FARM FLOCK USE

Ration 1 (grind at first; feed whole later)
Corn (no cob)	27.2
Oats	9.1
Wheat bran	4.5
Linseed or soybean meal	4.5
Trace mineralized salt	0.23
Bone meal or dicalcium phosphate	0.45
Alfalfa hay[3]	free-choice

Ration 2 (ground, pelleted, and self-fed)
Alfalfa, leafy ground	29.48
Soybean meal (44% CP)	4.5
Corn, shelled yellow	5.44
Oats, native white	4.08
Molasses	1.36
Bonemeal	0.45
Chlortetracycline, mg/kg	3.4
Vitamin A and D supplement	555 IU of A and 55 IU of D

Ration 3 (ground, pelleted, and self-fed)
Corn, ear	24.95
Alfalfa, leafy	13.61
Soybean meal	4.53
Molasses	2.27
Oxytetracycline, mg/kg of ration	3.4

Ration 4
Corn, ground	38.56
Soybean meal	6.8

Shelled corn and soybean pellets can be fed at same levels as above.
Ground ration can be hand-fed.
Whole grain ration may be self-fed.

RANGE USE

Barley, steam rolled	31.95
Dried beet pulp and molasses	13.61
Alfalfa hay	free-choice

1. Courtesy, National Academy of Science, Pub. No. 5, **Nutrient Requirements of Sheep.**
2. kg=2.2 pounds.
3. It is important to feed the best quality alfalfa hay in a separate rack. Be sure to feed hay and grain twice daily to keep it fresh.

If good quality clover, alfalfa, or cowpea hay is available, it may be used as the sole feed.

6. Using Antibiotics

More meat from less feed produced with less labor in a shorter time: This revolutionary claim is a scientific actuality with some classes of livestock. In recent years, a whole new area in feeding has been opened to scientists and farmers with the discovery that antibiotics can materially alter the feed requirements of livestock. The results to date have varied somewhat according to the class of livestock and the general health of the animals. Swine and meat birds, for example, have shown increased and more economical gains, and faster gains in many cases, as a result of including a small amount of antibiotics in their rations. However, dairy and beef calves have not been consistent in the effects they show from feed bolstered by antibiotics, unless the animals were sick or poor doers. Scouring calves are often helped by feeding small amounts of antibiotics. Young, growing animals respond to the substance whereas, to date, mature animals show little if any effect. There are many problems associated with feeding antibiotics. Mixing, for example, is extremely difficult, as it is hard to incorporate very small quantities of the antibiotic with feed so that it is evenly distributed. In addition, the whole science is so new that it is not justifiable to set down hard and fast rules until experiment stations and progressive farmers can examine the entire subject completely.

First feed and manage properly—Many favorable results from feeding antibiotics have come about when the livestock being tested have been "poor doers" or on improperly balanced rations. This statement does not intend to detract from the excellent results obtained from antibiotics; however, it should be pointed out that

Fig. 6.13—Lambs on full feed at an intermountain feedlot. A constant watchful eye must be kept on sheep in the feedlot to guard against over-eating trouble. (Courtesy, Union Pacific Railroad)

oftentimes the greatest need is simply sound management practices. If animals are well fed and managed, the need for additional aids is generally eliminated.

Do not feed antibiotics to sheep—Thus far, results on sheep have been too confusing to suggest their feeding as an approved practice. This idea is supported somewhat by the fact that antibiotics are expensive. With sheep and cattle, antibiotics do not produce results after the animal's digestive system is operating as a ruminant. For this reason it is logical to suspect that if antibiotics are used, they should be used when this class of livestock is young and their digestive systems functioning as a simple stomach like the stomach of swine. Carrying this logic one step further, it is even conceivable that antibiotics might harm ruminants because of the effect on the normal bacteria properly found in the stomach of ruminants. There may be a time when certain antibiotics or mixtures of them will be desirable to include in the ration, especially in the ration of sick sheep. However, in view of the cost of the product, it is recommended, according to present knowledge, that antibiotics not be included in the ration of sheep.

Keep posted—Conditions in the new science of antibiotics are being altered every day. Therefore, it is advisable to keep up on what is happening, as new discoveries may be to every sheepman's advantage. Your county agent, specialists in the universities, well-posted local veterinarians, farm magazines, and even good feed dealers are in a position to know what is new in the field. They should be consulted regularly, especially if your sheep are not making as efficient gains as they should.

7. Obtaining Good Buys

Ultimate success in the sheep business is determined largely by the amount of net profit a producer makes. Lowering feed costs is one way of increasing net profit. Therefore, proper purchasing of feed can mean the difference between profit or loss.

Make "off-season" purchases—Many items can be obtained cheaper by buying them when most ranchers are not ready for them and storing the materials until ready for use. Not only are supplies cheaper to obtain, but service and availability of materials are favorable to the purchaser. The price ratio is often so much in favor of "off-season" purchasers that it becomes a desirable practice to borrow money if necessary to finance these purchases. Alfalfa hay, for example, has considerable seasonal variation from summer to winter. Other items to save on include fuel, fertilizer and equipment.

Buy in large quantities—Piecemeal purchasing is

FEEDING AND FATTENING SHEEP

generally costly in the long run. Not only is the per unit price higher than in volume buying, but other factors, such as time and expense involved in numerous trips to the supply house, run up the total cost. Large-quantity buying results in better service from supplier, lower cost per unit plus the added advantage of always having enough feed and supplies on hand when they are needed. Large-quantity buying may make possible wholesale prices.

Buy direct from producer—Hay and grain purchased directly from the field are generally good buys. Handling and storage costs as well as middleman profits are eliminated by direct purchases. Even such items as lumber, when sizable amounts are needed, can be obtained at the mill at a considerable saving.

Watch for sales—It is surprising how many times during a year supplies and equipment will vary in price. Close-out sales, sales as a result of new models coming in, slack season sales, locality sales, sales as a result of being overstocked on certain items are all legitimate reasons that result in lower prices to the farmer if he will watch for them and take advantage of these bargains.

Plan your needs—Impulse buying is dangerous and often results in needless spending. Careful planning will result in only necessary items being purchased. It has the added advantage of knowing what you want so items can be purchased whenever a good buy on that particular article is offered.

Buy on T.D.N. basis—Purchasing feed on a basis of the **total digestible nutrients** in a bag of feed is the only safe way to buy many feeds. Oftentimes a higher price per ton really means a lower price per unit of T.D.N. Therefore, in most cases this would be the better buy. T.D.N. is written on the bag found on every sack of commercial feed. Comparative prices can easily be determined by figuring price per unit of T.D.N. according to this formula:

$$\frac{100}{\text{T.D.N.}} \times \frac{\text{selling price}}{\text{per 100 lbs.}} = \text{price per lb. of T.D.N.}$$

A comparison of the price per pound of T.D.N. for each feed is a desirable method of buying feed.

8. Sheeping Down for Fattening

Good gains are possible by field feeding corn and other crops even though it is, as yet, not a popular method in many parts of the country. It has the added advantage of utilizing grasses and weeds in fields that would go to waste as well as saving in labor.

Regulate corn at first—Ohio Bulletin 68 says that lambs turned into the cornfield should be allowed to eat a small amount of grain for the first week or so. It is a good idea to give them access to bluegrass or other pasture before turning them into corn permanently or some of them may eat too much corn. A full feed of hay will eliminate this possibility.

Supplement standing corn—Lambs do not make satisfactory gains on standing corn alone. Legume hay and .15 pound of linseed or cottonseed per lamb is best, although a full feed of legume hay alone in addition to corn is very satisfactory. Intercropping corn with rape or soybeans is an excellent and an economical way to balance corn that is meant to be field fed, according to **Farmers' Bulletin 1181**. In warm climates, cowpeas are also satisfactory. If neither legume hay nor pasture is available, about four ounces of finely ground limestone for every ten head of lambs should be provided.

Top corn stalks—Illinois Circular 523 states that some feeders cut off the cornstalks just above the ears so that the tops fall to the ground. This enables sheep to eat the upper portion as well as the lower part.

A small area may be cut each day.

Have pastures adjoining cornfields—Pastures of early seeded rye, wheat, winter barley, or stubble fields adjoining cornfields often provide an abundance of needed roughage to balance the diet.

9. Utilizing By-Products

In order to take full advantage of the grazing habits of sheep and their ability to convert waste products into meat, it is important to discover and utilize all by-products. The use of these foods in the sheep industry falls into two more or less separate avenues. First, small and large farms use by-products as a supplemental feed to hay and pasture, cottonseed cake, etc., to supply protein. Second, in areas where there are large numbers of fattening lambs around processing plants and where one by-product forms the basis of the fattening ration, the usual farm feeds are merely used to supplement and balance the by-product ration. Examples of this latter type are beet pulp, cottonseed cake or meal, cannery residues, orange pulp, distillery wastes, grape pumace, or any other by-product produced in ample quantities to encourage feeding of lambs on the site of production.

Search for new by-products—Every year new feeds are discovered or new uses found for old feeds. Almond hulls, for example, were considered a waste product on many California farms until the last several years. In one county alone (Butte), over 6000 tons annually are produced. Most of this valuable livestock feed was wasted until recent years and even now is not fully utilized. Sheep can handle almond hulls very efficiently. One can expect a pound of lamb for each four pounds of hulls and four pounds of alfalfa hay. If non-legume hay is used, about 10% of the ration should be cottonseed meal or soybean meal.

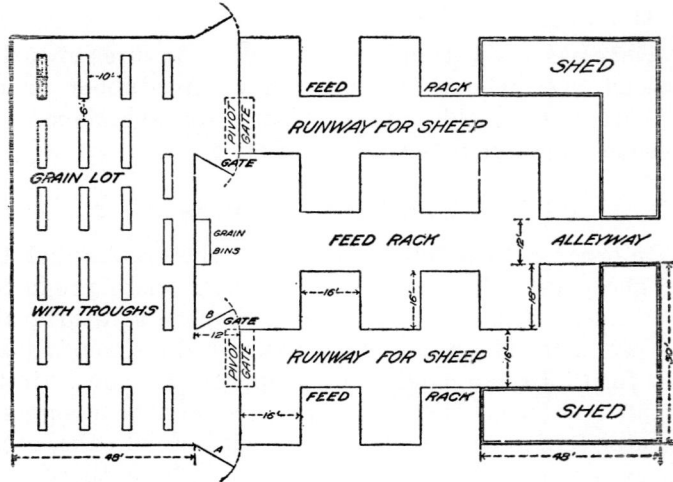

Fig. 6.14—Plan of a feed lot for finishing lambs for market.

According to **Cornell Bulletin 834**, ruminants are able to synthesize protein from non-protein nitrogen. Therefore, chemical products like **urea** can be used satisfactorily as a **partial** substitute or protein supplement. Lambs at the Cornell station fed urea as the **sole** source of protein did not make economical gains, however.

Molasses is another good by-product feed. It is high in carbohydrates and should only be used for reducing dust in mixed feeds and increasing palatability of most feeds. In some cases it has been sprayed on dry range grasses satisfactorily to improve their palatability.

Many other by-products exist, so it becomes a worthwhile practice to look around your community carefully and see whether or not some by-products exist that you can profitably utilize.

Make certain by-product rations are balanced—While excellent results are obtained from by-product feeding, oftentimes these feeds are critically lacking in certain elements. Almond hulls are very low in protein and

Vitamin A, beet pulp is very low in potassium, and cottonseed meal is low in vitamin A. Therefore, be sure to properly balance rations containing a large percentage of one feed.

10. Feeding Out by Contract

Frequently, in the areas where lambs are raised, there are not sufficient concentrates produced to fatten lambs properly unless they are milk fat and sent directly to market from their mothers. Under these conditions, growers may not realize full value for their lambs as they are sold for feeder prices. An additional factor favoring "feeding out by contract" is that your lambs are under the care and handling of a highly skilled feeder with all scientific feeding and disease control at his command.

Contract fattening of your lambs—This is a growing practice. Lambs are sent to a contract feeder who is located in a grain producing area, and the lambs are fed out on a specified price per pound of gain. There are other contract methods, but this is one common way. The farmer is benefited by getting fat price for his entire lamb even though he may just break even on the increase in weight from the contract feeder. Other contracts include: The **guaranteed-spread** contract which assures him a definite margin. The **feed-cost** contract gives the farmer specified prices for his feed plus an allowance for labor, use of equipment, etc. If lambs are fed on a **pounds gain** contract, a feeder is paid a stipulated price per pound for the increase in weight. A **cooperative share** contract provides for the division of receipts in proportion to the amount each party contributes.

Study the market—Contract feeding may not always be the best way of marketing, but if a seasonal rise in prices is expected, contract feeding will enable the

Fig. 6.15—A pen of show lambs. These lambs have been fed in such a manner that they have never been hungry since they were born.

rancher to hold his lambs 60 to 70 days before selling. He must consider that he also risks a lower price as well.

11. Feeding for Show

Lamb feeding for show varies little from good feeding practices used with commercial market production except that each practice is more rigidly adhered to and each factor must be at peak performance if a satisfactory show lamb is to be the result.

Keep lambs growing—Lambs must be kept gaining steadily in weight. This indicates good health and results in proper "bloom" to the lamb when ready for exhibition. Be careful not to force lambs too much during the beginning of the feeding period. Once lambs go seriously off feed, it is almost a lost cause to regain their bloom to meet stiff competition.

FEEDING AND FATTENING SHEEP

1960 CREEP FEEDING SUMMARY
AV. WEIGHT OF LAMBS AT 120 DAYS OF AGE

Lot 1 – Self-fed dehydrated alfalfa meal pellets (17% protein) throughout trial. — 83.10 lbs.

Lot 2 – Self-fed 17% alfalfa pellets throughout trial with shelled corn (self-fed) added at 56 days on feed. — 83.32 lbs.

Lot 3 – Self-fed 17% alfalfa pellets throughout trial with mix of 80% shelled corn – 20% soybean oil meal pellets (self-fed) added at 56 days on feed. — 88.55 lbs.

Lot 4 – No creep feed. — 59.03

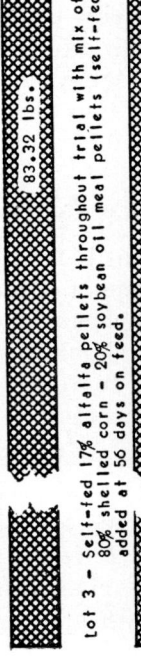

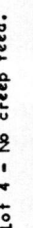

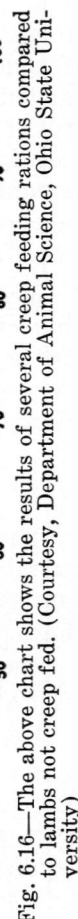

Fig. 6.16—The above chart shows the results of several creep feeding rations compared to lambs not creep fed. (Courtesy, Department of Animal Science, Ohio State University)

Creep feed—Even if lambs are running with their mothers, they should be creep fed. Lambs to be shown as spring lamb, of course, must receive their mother's milk as no other feed will equal gains made with this food. In addition to creep feeding grain, creeps into fresh pasture ahead of the mature ewes will insure added weight. Bran and oats, or oats and shelled corn, equal parts by weight, make a good ration for creep feeding. In hot weather, it may be wise to reduce the amount of grain and increase the green feed. If forage crops are not available, dried beet pulp is very satisfactory.

Keep ration high in protein—Almost any good fattening ration will put a finish on sheep, but it is usually better to have it a trifle higher in protein content than for commercial feeding. Use care to see that show sheep are not fed off their feet.

12. Creep Feeding Lambs

The creep is an area that is fenced so that the lambs can enter, but the ewes cannot. The openings in a creep are usually about eight inches wide and 15 to 18 inches high, but these openings should be adjustable.

Creep feeding is used to fatten lambs for market or to develop them for breeding.

Texas Agricultural Extension Service B-827 suggests that "A good developing feed to use in a creep for lambs is whole oats, and good fattening feeds are corn and sorghum grain. It is **not necessary to grind grain** for lambs since they chew their food well. Barley is also a popular feed with sheepmen although many prefer to feed it rolled."

Supply antibiotics for overeating disease—Antibiotics may be included in the creep ration to reduce scours and overeating disease. From 25 to 50 milligrams per pound of feed are recommended for suckling lambs while 10 milligrams per pound are recommended for older lambs. Aureomycin, terramycin and erythromycin are the most common antibiotics used in sheep rations. Follow manufacturer's directions.

Lambs usually start eating in a creep when only a few days old. Older lambs are more difficult to start in a creep.

13. Pelleting Feeds

Sheep eat pelleted feeds readily. Any kind of forage or grain may be pelleted either singly or in combination. The feeding value is generally different from plants fed unpelleted because the entire plant is incorporated into the pellet and, therefore, eaten by animals.

Normally animals would select more palatable portions, not eat stems or certain parts of plants so that the part actually swallowed would differ from pellets consumed.

The major disadvantage of pelleting is the cost of pelleting so this must be weighed against the advantages.

Important advantages are complete use of feed produced including little waste when fed. In addition, feeding operations can be mechanized and the benefits of bulk storage and handling utilized.

Pellet your own feed—In areas where pellet machines exist, home grown grains or forages can be used oftentimes at competitive costs with more conventional methods of harvesting feed. Even when costs are higher, over-all expenses may be less as feed is more economically utilized. Successful field pelleters are entering the picture, and this could further reduce costs. The decision is still one which each producer must make depending upon local costs of harvesting.

See **Approved Practices in Feeds and Feeding.** A detailed feeding program can be found in this book, which is available from The Interstate Printers & Publishers, Inc.

Feed Cubes—Sheep eat cubes readily.

CHAPTER VII

SHELTER AND EQUIPMENT FOR SHEEP

Proper shelter, equipment, and handling facilities can easily make the difference between profit and loss in the sheep industry. In particular, ease of handling and reduction in number of man hours to handle and care for sheep can be materially reduced by utilizing good, common sense handling facilities designed to reduce the necessary amount of labor. Proper feed racks, watering facilities, and the like can also effect a material saving in feed costs as well as keep areas where sheep congregate, clean and germ free. However, it is entirely unnecessary that fancy buildings or high-priced equipment be used in order to grow healthy sheep. Newborn lambs must be kept warm in cold weather, but ewes need only to be kept dry. Plenty of ventilation and light should be available at all times.

In order to care for sheep properly and yet maintain capital investment at the lowest, safe level, it is necessary that a number of approved practices be followed.

Activities Which Involve Approved Practices

1. Protecting small farm flocks.
2. Protecting range sheep.
3. Getting proper arrangement.

Fig. 7.1—Proper equipment is necessary for success in the sheep business. A good pickup truck fitted with racks for hauling sheep is a must on every ranch. Notice the simple, sturdy construction used on this rack. Bolts and metal braces are used for strength and lasting while the wood rails have been planed on the edge to prevent splintering and catching on the wool.

4. Obtaining proper feeding equipment.
5. Obtaining proper watering facilities.
6. Maintaining adequate fencing and corrals.
7. Designing specialized equipment.
8. Miscellaneous equipment.
9. Providing correct loading and shipping facilities.
10. Providing suitable feed storage.

SHELTER AND EQUIPMENT FOR SHEEP

1. Protecting Small Farm Flocks

Under mild winter conditions, little protection, other than a windbreak and an area where ewes can get under cover during rains, is necessary; however, under winter conditions that exist in many parts of the country, adequate protection is desirable both from the standpoint of the operator and the health of his stock.

Utilize sections of present facilities—There is little need for extra construction on most farms having small flocks. Sections of barns or storage facilities can be fenced off inside often with temporary panels that will do very well for protecting sheep. Such facilities are often already well-lighted and ventilated so that a minimum of effort is necessary to convert them for use. Once the flock reaches say 100 ewes, then separate buildings should be considered.

Maintain adequate light and ventilation—Sheep shut up in dark, damp houses won't thrive. Such facilities breed parasites and the animals cannot dry out even

Fig. 7.2—A well-ventilated section of a barn used to house a farm flock. Facilities for sheep need not be expensive.

from moisture produced just from their own bodies. Windows or shutters should be so arranged that they can be opened wide whenever weather permits in order for floors and walls to dry completely. Roof ventilators are desirable in permanently constructed sheep barns, as they continually let out warm, damp air and permit circulation.

Choose correct location—Permanent buildings for sheep should be placed where it is dry and well-drained. In addition, dry and sheltered yard space should be available adjacent to the main barn or shed. A southern slope with sandy soil is especially suited for this purpose. It is also advantageous to have the buildings and pens convenient to the main farm house and easily reached from regular pastures and fields.

Use temporary structures—Oftentimes bales of straw stacked to form the walls and loose boards placed on top with tarpaper and more straw can be utilized as inexpensive shelters for sheep. Poles placed in the ground, then covered with straw after stretching fencing over the top, can also be used to great advantage.

Keep dark in summer—Structures designed so they will be quite dark during the summer months have an advantage in keeping flies away from sheep during the heat of the day. Straw covered, temporary structures will do very well for such purposes.

Use dirt floors—Dirt, preferably tightly-packed clay, makes the best floors for sheep barns, according to **Kansas Bulletin 316.** Concrete floors for alleys and feed rooms are necessary in order to save feed and keep out rats, but such floors are unnecessary in the pens.

Pen partitions should be movable—In this fashion the size of pen space can be varied as needed or removed in case the facilities are to be used for other stock. Cleaning, also, is speeded up as the partitions can be removed and mechanical equipment utilized to speed up the operation.

Use wide doors and low sills—According to **Kansas Bulletin 316,** sheep will not crowd in a wide doorway, whereas crowding through a narrow doorway might cause injury. High sills are particularly dangerous to pregnant ewes as they have to jump over them.

2. Protecting Range Sheep

Many of the ideas used in penning and housing farm flocks can be readily transferred to the handling of range sheep. However, large bands often need special facilities designed especially for them and not made-over equipment as frequently is the case with farm flocks.

Build for ease of shepherd—It is a mistaken idea that sheep barns must be built to fit just the sheep. While it is true that unnecessarily high sheds, corrals, etc., cost more and may be too cold in winter, sheep sheds and barns should be high enough so that any herder or operator can enter easily without stooping or crouching.

Plan for storage—With range sheep this is an important item as large amounts of feed may suddenly

Fig. 7.3—An open shed with feed racks on wall.

become necessary. Farm flocks on the other hand can always get by with a little extra effort on the part of the owner. The wise range operator generally has on hand more than enough feed to meet an emergency such as a prolonged snowstorm or blizzard.

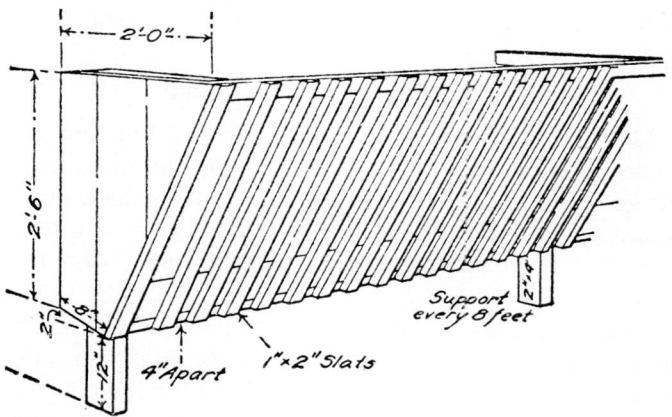

Fig. 7.4—This feed rack attaches to the wall, thereby saving valuable space.

Select sheltered sites—All ranges have some areas that are more exposed than others. Very often simply observing where livestock like to bed down and relax will give one a key as to where a good building site is located. Nearness to water, relationship to existing fence lines, and availability of all-year roads are important items to consider in addition to locating the sheds and barns in naturally protected areas.

Use native materials—In almost all cases, materials on hand and easily available will do nicely for use in constructing barns, sheds, corrals, corner fences, and the like suitable to house and protect sheep. Native stone, poles, sod or adobe bricks, shakes for roofing, brush for fences, lumber cut from native timber, thus eliminating expensive hauling and freight rates, will all be satisfactory as construction materials.

Use modern bracing and preservatives—Many sheep structures are thrown together so hastily that labor and materials are not used to full advantage and buildings fall down or are in need of repair long before it should be necessary. New type materials like inexpensive steel bracing for rafters, sills, modern electrical outlets, and plumbing are available. These materials are superior building products, and reduce time and labor in construction. Boards that rest on the ground should be of rot-resisting material such as redwood or, better yet, parts that contact the ground should be made of concrete. Preservatives like creosote or penta products should be copiously applied to all lower-wood members of a building or fence, especially when poles and unsawed materials are used in construction.

One or two threaded steel rods used in connection with turnbuckles will securely brace a building or fence and can easily be tightened when the lumber dries out or buildings become old and start to sag.

3. Getting Proper Arrangement

Time and motion studies have shown that many farmers walk countless miles and perform dozens of unnecessary chores, such as opening gates that could be eliminated or prevented by careful planning and thought given to the arrangement of corrals, sheds and interior partitions, doors, etc., in buildings.

Plan for minimum waste space—Many buildings have dark, unused corners that could be eliminated or made useful by the addition of a door or window. Great, high ceilings often go unused whereas the addition of a loft would give valuable loft storage for feed or for rarely used equipment. Hoists rigged from the rafters will get

Fig. 7.5—Proper arrangement is important. Watering facilities are arranged in the above illustration so sheep can water from either side of the fence. Gates, too, are located in the corners of fields, facilitating easy movement.

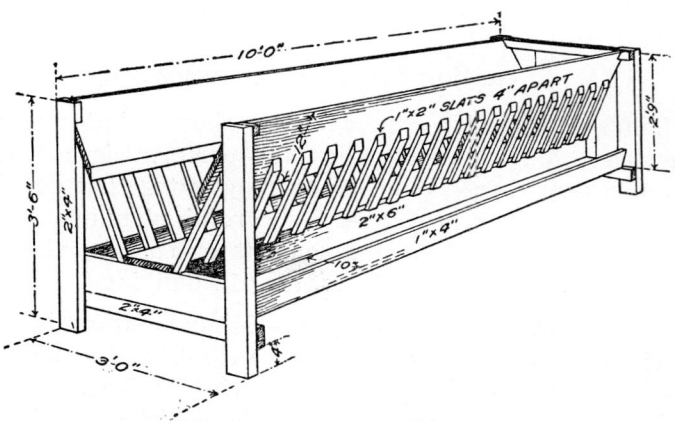

Fig. 7.6—A combination hay and grain rack which may be entered by attendant when feeding grain. This rack may also be used as a pen partition.

panels and other bulky items out of the way so the entire ground floor can be used for sheep.

Occasionally corrals are made too large for the number of animals to be handled so that needless fence repair and construction are necessary. In addition, livestock cannot be handled and sorted efficiently, etc., in corrals unnecessarily large. Ordinarily, 10 to 12 square feet per animal is ample for a corral or pen.

Design for ease of cleaning—Doorways should be wide enough for tractors, manure spreaders, or wagons to enter. If such an arrangement is unavailable, then some easily removable section of the wall should be incorporated so that power cleaning is possible. Inside pens and partitions should be large enough, or else removable, so that cleanings can be accomplished with a minimum of hand labor.

Utilize movable pen partitions—Farmers' Bulletin 810 suggests that pen partitions should be movable. By using feed racks to make divisions in the pen space, the size of the pen can be varied as needed, and in special cases the racks can be removed to permit the use of space for other stock.

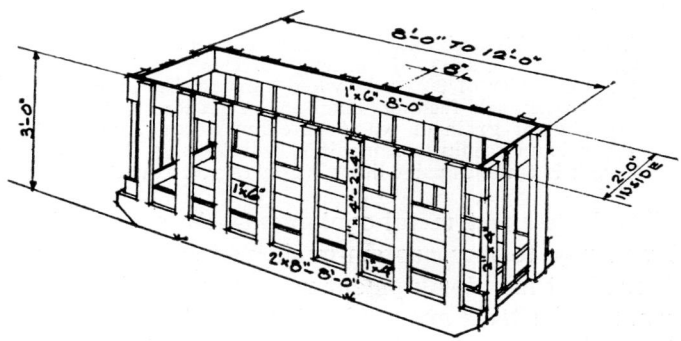

Fig. 7.7—Hay and grain rack for sheep.

4. Obtaining Proper Feeding Equipment

It is necessary that feeding equipment be the right size so that sheep do not overcrowd, that it be of the right design so that feed is not wasted and that it be economical so that a minimum of money is invested in order to show a high net income in the sheep business.

FEED TROUGH

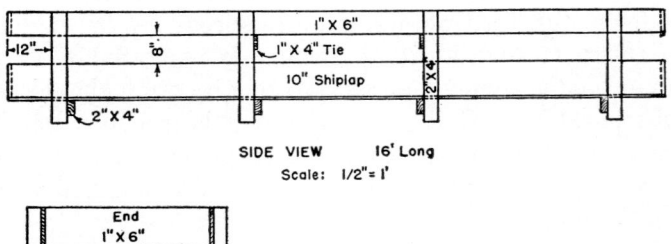

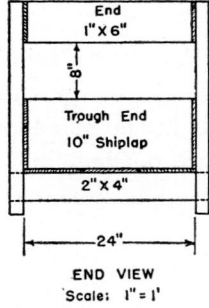

Fig. 7.8—A feed trough may be built any convenient length; 10, 12, 14, and 16 feet are common lengths. This combination feeder has a capacity of 30 lambs when hand feeding, and 60 to 80 lambs when self-feeding. This kind of feeder can be used for any kind of grain or whole or chopped roughages and for either hand feeding or modified self-feeding. To clean, the trough is turned upside down. (Courtesy, Texas A. & M.)

Purchase sheet metal equipment—As a general rule it will pay to buy custom-made watering troughs or other sheet metal equipment rather than building your own. Special tools are needed in order to properly construct sheet metal equipment, and a certain amount of

SHELTER AND EQUIPMENT FOR SHEEP

skill and practice in order to make smooth, strong, nonleaking equipment. For this reason it is a desired practice for most operators to purchase ready-made sheet metal equipment.

Use good grade, smooth lumber—It is poor economy to attempt to get by with low-grade lumber. In the first place, a relatively little amount of lumber is needed to build most feeding equipment. Cost of labor is relatively high and good lumber makes more efficient use of labor. One weak board can often ruin an entire trough and necessitate a lengthy repair. One x three slats and 2x4 or 2x6 are common sizes required for most feeder construction. It is advisable to use all surfaced lumber as the wool catches easily on rough sawn lumber.

Consult plans before building—Time and effort are saved and there is less waste of material when a good plan is used. Many plans are available in **Farmers' Bulletin 810**, or from the county extension office as well as commercial companies interested in sheep. These plans are free and will result in smoother, faster

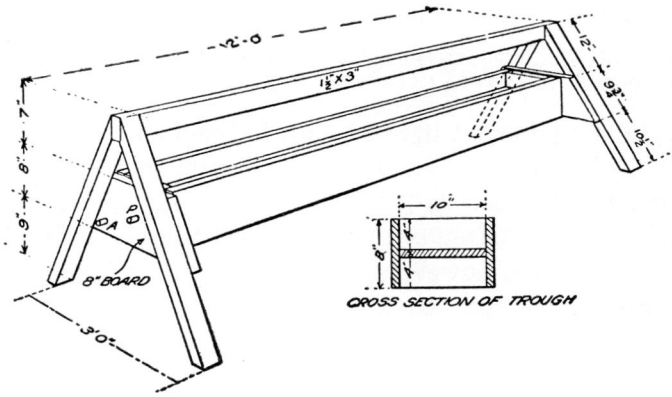

Fig. 7.9—A reversible, movable grain trough.

construction of feeding equipment. Width necessary per ewe, height from ground, length of feeders, panels for ease of handling, and other factors are all carefully figured out thus saving the farmer the trouble of making his own plans. If no plans are available, it is well to work out a simple drawing on paper before beginning construction. Combination hay and grain racks constructed properly generally allow one and one-half to two feet of trough space per ewe. Slats should be close enough on hay racks to prevent ewes sticking their heads through.

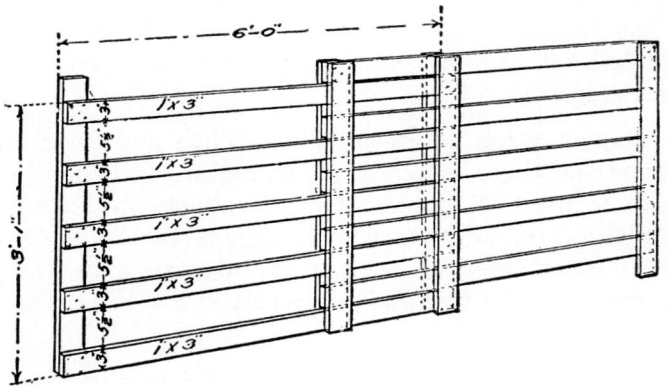

Fig. 7.10—A light extension hurdle. May be closed to 6 feet 4 inches or extended to 11 feet 4 inches.

Build equipment light enough to handle easily—It is a mistake to build equipment so heavy that one man cannot easily handle it. This is especially true during wet weather. Often equipment is made heavy so it will not turn over; however, if it is properly designed and braced, even light equipment will stand considerable abuse and be relatively stable. Proper use of braces and triangles in construction will keep equipment strong and true to shape.

SHELTER AND EQUIPMENT FOR SHEEP

Use bolts and metal braces—One or two strategically placed bolts will do much more to hold equipment together than many nails or extra bracing. Bolts should be of carriage or smooth-head design and used with one flat washer. The length should be only long enough to come flush with the end of the nut and not have to be filed or sawed off as wool will catch on the jagged edges. One-fourth or $\frac{5}{16}$ inch diameter bolts are suitable for most construction.

Corner braces and strap metal can often be used to strengthen crates or joints getting severe use. Such devices are excellent because they not only hold securely but spread the strain over a greater area of the wood.

Use concrete for permanent installations—Feed yards, water troughs, development of springs, and similar installations are best made of concrete when long time use is considered. Concrete aprons around water troughs are desirable as they tend to keep the area clean and dry and, therefore, freer of parasites. Paved yards are easier to clean, particularly when using mechanical equipment.

5. Obtaining Proper Watering Facilities

Water is an important item to consider in designing facilities for raising sheep although not so vital as it is with other classes of livestock especially during win-

Fig. 7.11—Range water supply for sheep: (A) concrete storage tank built into slope of watershed; (B) collection ditch diverting snow water run-off into storage tank; (C) control valve; (D) pipe line; (E) trough.

ter months. Range sheep are irregular in their watering habits depending, of course, upon time of year, type of vegetation being eaten, climate, etc.

Develop natural water—Springs can be cleaned out, rocked up, and piped to troughs so that clear, drinkable water is often obtainable from otherwise useless sites. Even seepage areas and natural run-off frequently can be impounded with little effort and expense into a surprisingly good source of water. The local soil conservation service will give valuable aid in determining locations and how such devices should be constructed.

Build for ease of cleaning—Portable or permanent watering facilities should all be designed so they can be easily cleaned out and scrubbed, if necessary. Dirt and growth collect inside, often harboring internal parasites, so it becomes imperative that they be cleaned. Most watering troughs are partly covered. These coverings should be installed so they can be lifted off for ease of cleaning.

Use automatic water valves—Time and labor are saved by the use of automatic valves and, in addition, there is always an ample supply of water guaranteed. Such valves are inexpensive and relatively trouble-free if they are protected against mistreatment by livestock. On ranges where windmills or springs provide a continuous flow of water, it is desirable to have a series of troughs to catch excess water or else pipe the excess far enough away so the ground around the trough does not become boggy. Wet areas that sheep have to stand around in are danger zones.

6. Maintaining Adequate Fencing and Corrals

In the past when sheep raising consisted mainly of grazing on the open range, fencing was entirely unnecessary. With little open range left and with huge numbers of small farm flocks, fencing is, at present, one

SHELTER AND EQUIPMENT FOR SHEEP

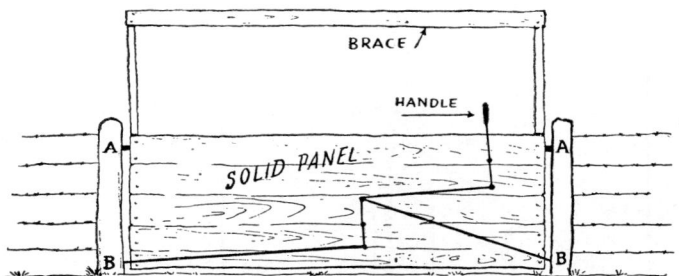

Fig. 7.12—Pivot gate—This gate pivots at points A-A, and latches at points B-B. In this fashion it can open in either direction, permitting sheep to pass underneath.

of the major cost factors in the sheep industry. In addition, ease of handling, efficiency of operation, and time saved are dependent upon a good system of corrals and handling facilities. Even large range breeders have a system of corrals on their ranches designed to speed up handling and reduce labor.

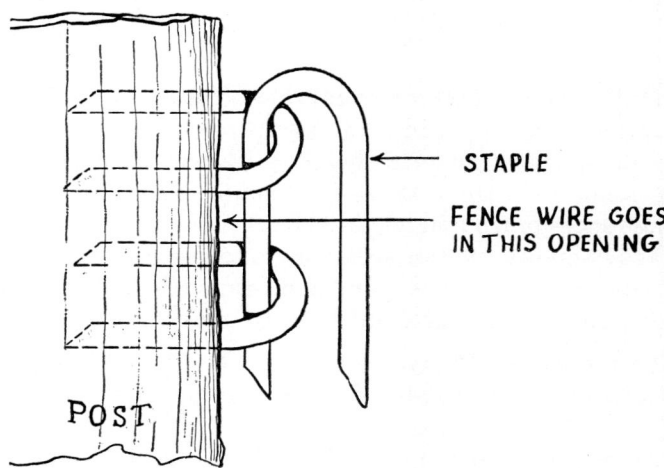

Fig. 7.13—Where fences must be taken down frequently, a demountable fence fastener can be made with three staples. Lift free swinging staple to remove fence and replace as indicated so staple will not be lost.

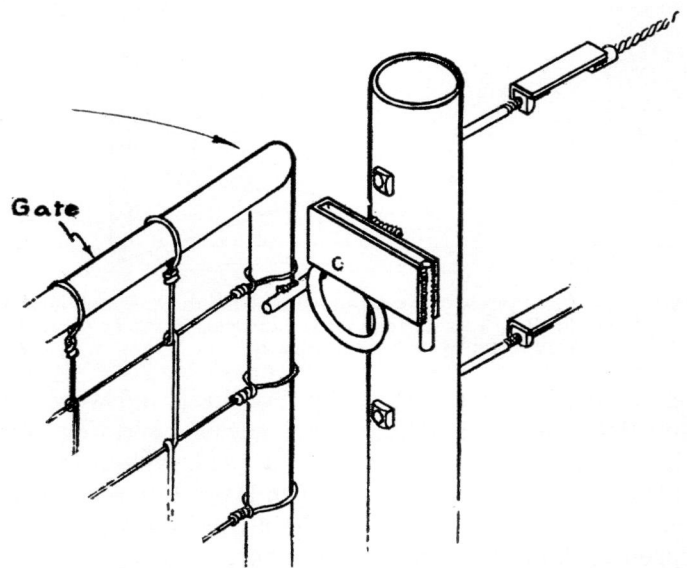

Fig. 7.14—Ring latch—Illustrating single latch for metal gate. Gate shown being closed.

Make snow fences demountable—Three staples, two of which are driven into the post, like the accompanying illustration, make a cheap, quick way of demounting fences. In some areas it is sound management to take fences down and let them lie flat during the winter if heavy snows tend to break down fences during a part of the year when they are not being used. Irrigated pastures also are better utilized by movable fences.

Plan fields carefully—Small pens or lots with access to shelter are invaluable during lambing season. With farm flocks, they should also be near at hand to the farm home so shepherding becomes an easier job. On the other hand, too many small fields give a cluttered-up appearance to a farmstead and may actually interfere with other farming operations. In order to fit in

SHELTER AND EQUIPMENT FOR SHEEP

with growing crops that require plowing and harvesting equipment, it is doubtful if fields smaller than 10 acres are economical. Fence rows become dead spots on a farm, not to mention the fact that fencing and fencing maintenance are expensive. All fields should have gates large enough to permit the largest equipment one has, or will need, to enter.

Electric fences have been used with only moderate success although as a means of dividing irrigated pastures into grazing units, they are rather common. It will be necessary to fence irrigated pastures into smaller units than dry range.

Locate gates in corners—It is difficult to herd sheep through gates unless they are located in natural turning points or corners. This usually makes a more

Fig. 7.15—Well constructed pens, gates, and equipment used to handle sheep. (Courtesy, Union Pacific Railroad)

sensible arrangement from the standpoint of going into other pastures either with sheep or equipment.

Use solid materials for corrals—Wire is not desirable for corrals as sheep tend to catch their fleeces in the joints whenever crowded. They also run into it whenever they become panicky. Most sheepmen prefer 1x6 lumber, although concrete walls, brush fences, or small poles and rails are entirely satisfactory if well constructed. Four-foot high corrals are adequate for sheep unless dogs are a problem or other livestock are to use them.

Consider using hurdles—Portable hurdles or panels are sometimes used to graze pastures such as rape or other forage crops. One novel idea for using hurdles is an iron hurdle which will retain the ewes but permit the smaller lambs to pass through. This enables the lambs to get better feed than their mothers. The hurdles are moved forward as the ewes clean up the feed behind the hurdle.

Build a cutting chute—Every sheep ranch should have a well-made cutting chute useful in sorting and grading sheep. Much time and effort will be saved by such a device. Cutting chutes should be around 40 inches high, made of smooth 1x6 lumber, with fence-

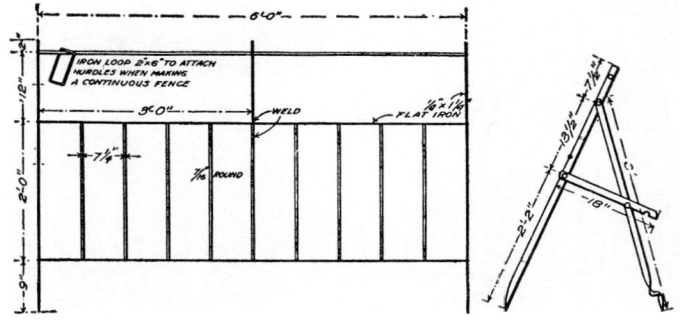

Fig. 7.16—An iron hurdle used in close grazing of forage crops. (Courtesy, USDA)

SHELTER AND EQUIPMENT FOR SHEEP

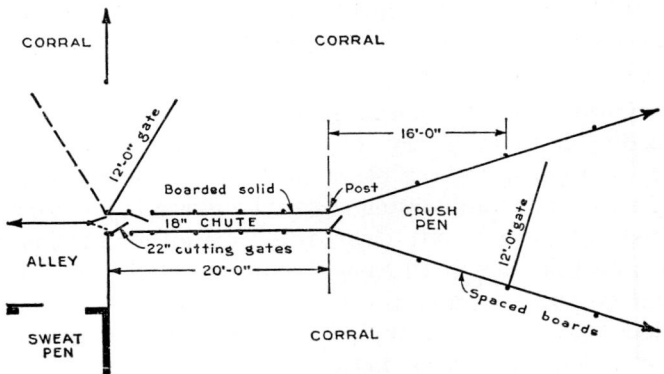

Fig. 7.17—Plan of cutting chute. (Courtesy, USDA)

posts on the outside, approximately 18 inches wide (inside measurement), sturdily constructed, and with a strategically located "dodge" gate.

7. Designing Specialized Equipment

There is a large number of different types of equipment designed to make sheep raising easier and oftentimes more profitable. Some of this equipment is excellent; other ideas may prove valueless. Each grower

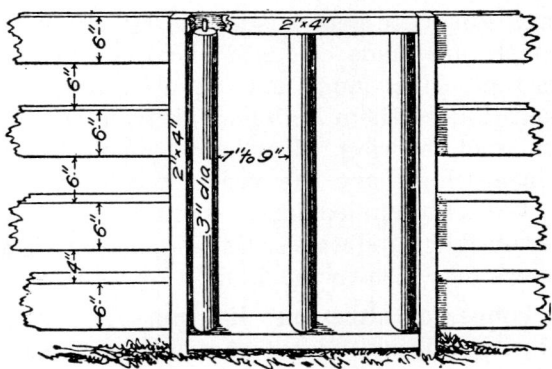

Fig. 7.18—Lamb creep with rollers for uprights. Large lambs can squeeze through.

should carefully size up his own farm and its needs to see what kind of specialized equipment will work out best in his situation.

Creeps—Range men make little use of creeps. Creep feeding is unnecessary if good, well-fertilized pastures are available. On the other hand, with show lambs, if ewes are poor milkers or the range grass is poor, creep feeding may pay off. Creeps may be either stationary or movable, but in all cases are designed so that lambs can go through and the ewes cannot. Rollers on the sides may improve a creep but are not absolutely necessary. The distances between slats forming a creep should be seven to nine inches. Very smooth boards, pipes or rollers may be used, as rough lumber would tear at the wool.

Brooders—Once a lamb is dry and its belly full, it can stand a lot of hardship. Brooders can help in this respect and may prove very useful during foul weather. Electric brooders are the most practical, although running hot water in pipes from a central lambing house is a desirable practice. Generally, ewes are kept in square pens 50 to 60 inches on a side, and the brooder placed in one corner about 12 inches off the bedding. Adequate heat is supplied by 150 watt bulbs if brooders are made of 1"x12" boards and have aluminum tops. Heat lamps may be used, but care must be taken to keep them high enough from the lambs to prevent wool charring. Heat lamps should be hung 42 to 48 inches high, and the ewe fenced away from the lamp. Weak or chilled lambs should be placed in a brooder until dry and warm. Usually, one day is enough as the ewe may disown the lamb if kept away too long.

Sun lamps for ultraviolet lighting—Young animals born during fall and winter and confined indoors may miss the helpful benefits of sunlight. Ultraviolet light from sunlamps may enable them to build up sufficient

SPACE REQUIREMENTS OF BUILDINGS AND EQUIPMENT FOR SHEEP

Class, age and size of animals	BARN OR SHED			FEED LOT		HAY RACK				FEED TROUGH (For Grain, Roots or Silage)				SELF-FEEDER	Water per animal per day, (gals.)
	Floor area per animal (sq. ft.)	Height of ceiling, (ft.)	Window space (not incl. open sheds) (sq. ft.)	Area if ordinary dirt lot, (sq. ft.)	Area if paved lot (sq. ft.)	Length per animal (in.)	Width if feeds from 1 side (in.)	Width if feeds both sides, (in.)	Height at throat (in.)	Length per animal, (in.)	Width if feeds from 1 side (in.)	Width if feeds from 2 sides, (in.)	Height at throat, (in.)	Trough length if feeder is kept filled, (in.)	
Dry ewes	12-20	8½-10	1 sq. ft. window space per 35 sq. ft. floor space	25-40	20-30	18-24	14-16	20-24	12-15	16-22	14-16	20-24	10-15	....	1½
Ewes with lambs	15-22	"	"	25-50	25-40	18-24	"	"	"	"	"	"	"	....	2 "
Stud rams	20-30	"	"	30-60	"	"	"	"	"	20-24	"	"	"	....	" "
Feeder lambs	10-12	"	"	20-30	10-20	12-15	12-14	18-22	10-12	12-15	"	16-18	8-12	3 lambs for each 12" feeder space	½

From Circular 314, Washington State Agricultural Experiment Station

Remarks: Wide barn doors are needed to prevent crowding and possible injury to pregnant ewes. Doors at least 8 feet wide are preferable.

Vitamin D content in their bodies. This is not a general practice, but in areas where lack of sunlight is a problem, it may prove to be advantageous.

Scales—It is true that a good sheepman has an "eye" for his animals which enables him to tell at a glance their general health, whether or not they are "doing well," etc. However, even these folks find a weighing scale an invaluable piece of equipment in their over-all livestock program. Obviously, the sale of livestock on a ranch is greatly facilitated by the availability of a good scale. In addition, ranchers will gain because "shrink" will not be as great if livestock are sold on the home ranch and weighed there, even though buyers will take the decreased shrinkage factor into consideration. Sheepmen have a better idea of what to expect at market if they weigh animals before leaving the farm and can also determine accurately the shrinkage by comparing these weights with terminal selling point weights. Probably the most important advantage of having a good, home scale arises from the value one obtains from periodically weighing random numbers of lambs or feeders to determine the rate of gain while fattening sheep. By this practice, feed requirements can be determined more accurately and fat stock shipped at their peak, bloom stage, and most favorable market weights.

Scales should be located close to loading and shipping facilities and designed to fit into the over-all corral and handling chute arrangement so that weighing does not become a difficult task. If sheep are the only livestock on the farm, a small scale designed specifically for sheep can handle the job nicely. On the other hand, if the rancher has other livestock or plans to have larger animals, it is wise economy to construct scales and chutes large enough to handle mature beef animals. Scales should be sturdily built, with sides high enough

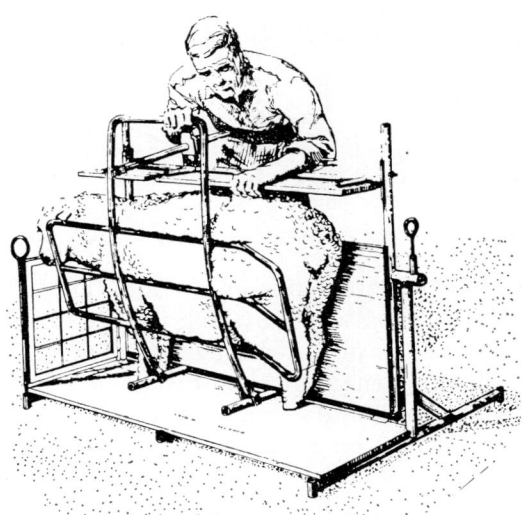

Fig. 7.19—Tilting squeeze for sheep. This will enable easy inspection, examination, and treatment, especially for a single man operation. (USDA Plan 6006)

to prevent animals from jumping over (about six foot sides), and covered by a roof to give protection during rainy weather. Small sheep breeders can get along very well by simply utilizing a shipping crate large enough for one animal and placing it upon an ordinary farm grain scales. Many boys and girls starting out in the sheep business with one or more show animals will find this an entirely satisfactory practice. In the absence of a crate, a lamb can be held by a person and its weight determined by subtracting the weight of the person from the combined weights to find the weight of the lamb.

Dips—A dipping vat makes it possible to thoroughly and easily treat various skin troubles as well as assisting in preventing lice, ticks, and other external parasites. Many range sheepmen make it a practice to periodically dip their flocks as a sound preventive

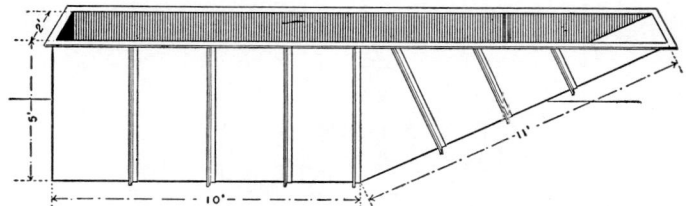

Fig. 7.20—Portable galvanized-iron sheep dipping vat.

measure. Whether one dips or not and how often will depend to a large extent on which part of the country one lives in and the prevalence of parasites. (See Chapter VIII.) Many owners of farm flocks have no dipping vat, but if the flock is to become a permanent enterprise on the farm, it may be advisable to install a simple vat. Different types of vats are available. They may be constructed of redwood, concrete, or steel. **Farmers' Bulletin 798** has plans and full descriptions of a great variety of dipping vats. One very simple method, according to **Kansas Bulletin 316,** is to purchase a galvanized iron tank of proper size and shape which is then set into the ground. A wooden runway can be built to the deep end of the vat. A small holding pen is constructed at the other end so sheep can be held until the solution drains from them and back into the vat. The holding pen is essential, as sheep carry large amounts of fluid with them from the tank and if this is not allowed to drain back, the tank must be replenished too often. Nicotine sulfate is a common dip although many standard dips are available at livestock supply houses. Portable dipping vats made of canvas are available should one desire to dip only a few animals. This apparatus is valuable if dipping is to be done in scattered areas. It is made of heavy canvas (#40) as follows: Two strips of canvas 8 feet long and 26 inches wide are sewn together to form a bag 48 inches deep and 96 inches in circumference. Seams are triple sewed

SHELTER AND EQUIPMENT FOR SHEEP

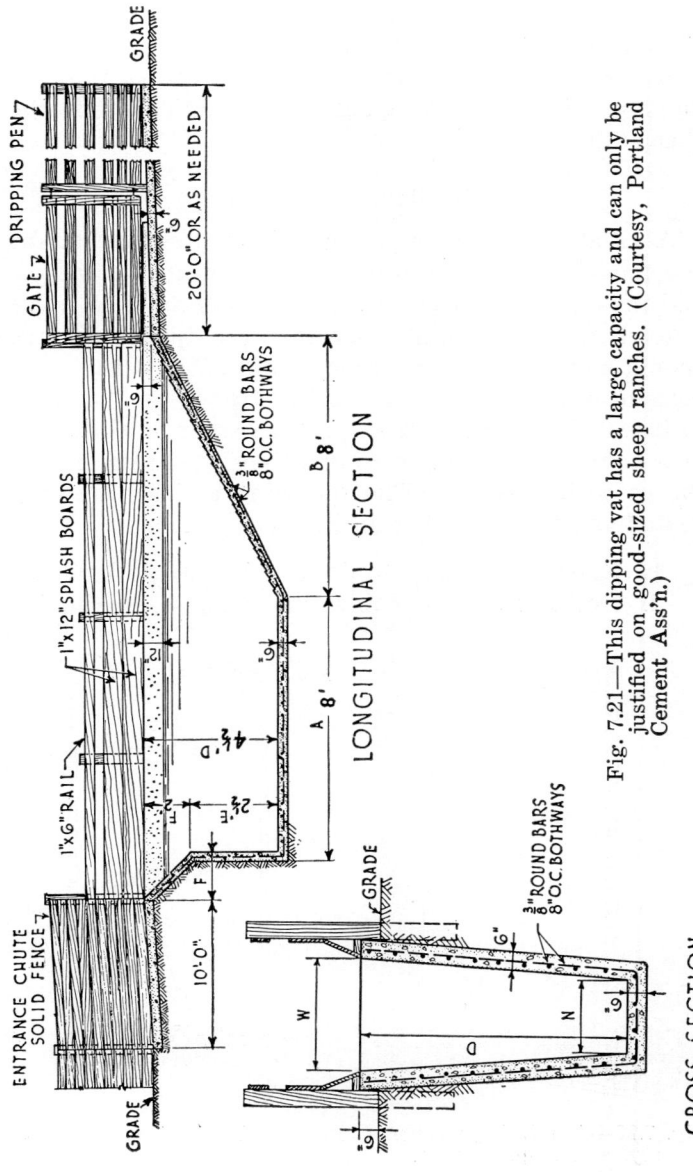

Fig. 7.21—This dipping vat has a large capacity and can only be justified on good-sized sheep ranches. (Courtesy, Portland Cement Ass'n.)

and tops and corners reinforced with leather. Iron rings are riveted to the upper part of the bag so that it can be supported while animals are lowered into the dipping solution. It is a desirable practice to tie the sheep's feet first.

8. Miscellaneous Equipment

A great variety of equipment is on the market for sheepmen to consider in order to facilitate operations and make their job easier. Much equipment is home constructed and is quite satisfactory. Livestock men should not invest nor become burdened with every piece of equipment on the market. However, it is wise to investigate each item and see whether or not it fits into your particular farm operations and then obtain the essential articles. The following articles are suggested as being useful on most sheep ranches:

1. Elastrator—used to dock and castrate lambs.
2. Docking stool—enables one man to do the job

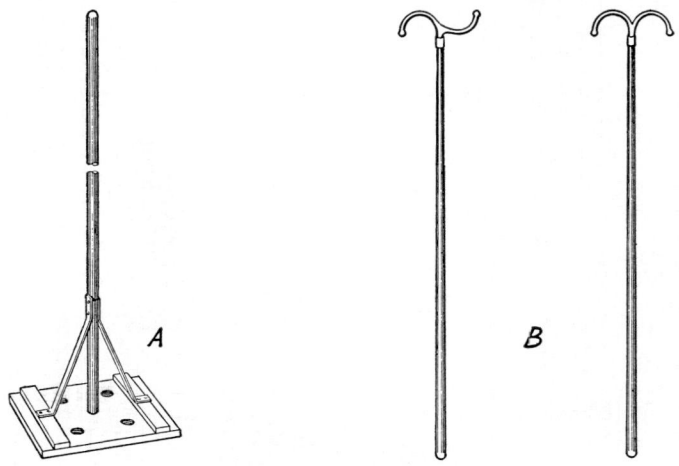

Fig. 7.22—A, stirring plunger for mixing liquids in vat; B, two styles of dipping forks.

without burning lamb, as well as insuring uniform length docks but is unnecessary if the elastrator is used.
3. Docking chisel—made from 1¼ inch gas pipe.
4. Pruning knife—for trimming feet.
5. Pruning shears (long handle)—excellent for trimming feet when grown out too long or in dry weather.
6. Punch—used to punch ears to insert metal tags.
7. Ear notcher—used to notch ears as a means of identification.
8. Tattoo outfit—the surest method of identification—recommended for purebreds.
9. Three-ounce, hard rubber syringe—useful in making injections and cleaning wounds.
10. Drenching tube—useful in administering copper sulfate drench. It consists of a 4-foot piece of rubber tubing, a granite-ware funnel, and a 6 inch piece of ⅜ inch brass pipe.
11. Graduated nursing bottle and nipple—for measuring medicines and raising orphan lambs.
12. The medicine chest—**Farmers' Bulletin 810** says, "The following medicinal equipment should be available at all times. The quantities mentioned are for the customary farm flock of from 40 to 50 ewes. In case larger flocks are kept, the amounts should be correspondingly increased."

Compound cresol solution or an equivalent preparation. (This may be used in a 2 per cent strength for the washing of wounds and in a 3 per cent strength for the disinfecting of pens.)	1 gallon
Epsom salt	5 pounds
Castor oil	1 quart
Copper sulfate (blue vitriol)	5 pounds
Boric acid (same as boracic)	1 pound
Tincture of iodine (in a glass-stoppered bottle)	½ pint
Lime	100 pounds
Absorbent cotton	2 pounds
Two-inch-wide muslin bandages	1 dozen

Much of the miscellaneous equipment is available commercially, and it is a desirable practice to purchase this equipment from livestock supply firms.

9. Providing Correct Loading and Shipping Facilities

Loading of animals, even with small lots, is greatly facilitated by the use of proper shipping facilities. In addition, there is less chance of injury to livestock and humans.

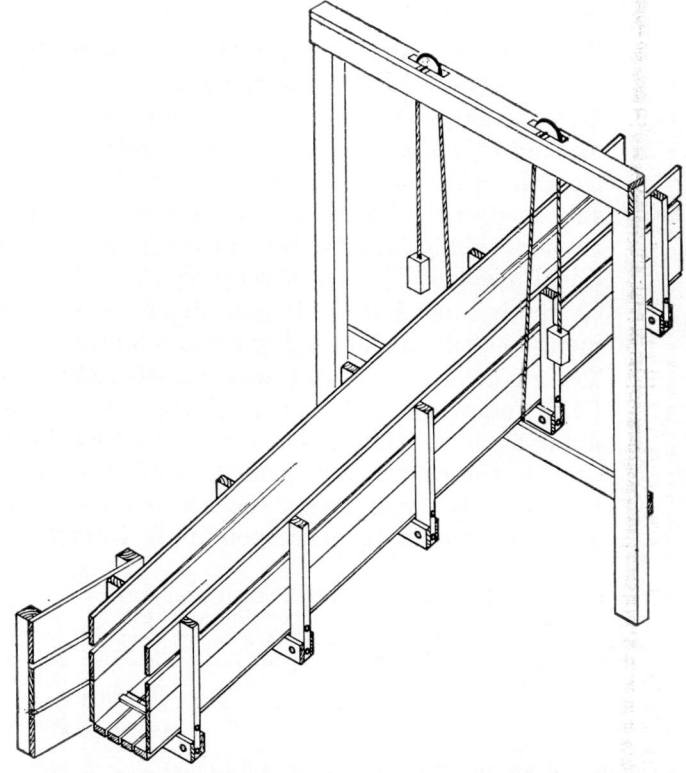

Fig. 7.23—Loading chute for sheep. (Complete plans are available from Agricultural Publications, University of Calif., Berkeley, Plan No. 75.)

Use proper dimensions—Frequently money is wasted because expensive materials are carelessly thrown together for a makeshift handling system. Height of ramp may be 36 to 48 inches depending upon height of truck to be used. Slopes may vary but should not be steeper than one foot high to six feet horizontally. The width is most important. A recommended inside width for a loading or cutting chute is 22 inches. Dodge gates should be around 32 inches long and 40 inches high. If only sheep are to be loaded, 48 inches at the truck end is sufficient.

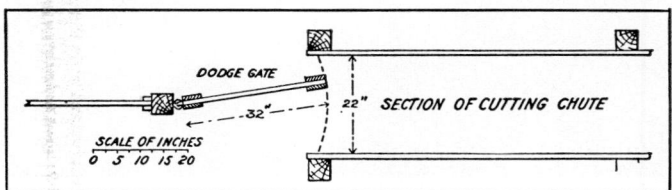

Fig. 7.24—Dodge gate located at the end of chute can quickly separate sheep whenever necessary.

Use solid construction—Most any material may be used, but stout 4x4 posts set well in the ground or solid concrete is preferred. Animals hesitate and often refuse to walk on shaky floors or equipment. Therefore, it is extremely important to design loading ramps solidly. Dirt corrals are excellent unless large numbers are to be handled in wet weather. Then asphalt or bitumuls type pavements are recommended because they are not as slippery as concrete. Redwood bark for runways or solid redwood planking is excellent although somewhat costly. Such flooring might be justified for commercial feedlots especially in damp climates. If concrete is used, it should be roughened.

Consider multiple-purpose ramps—If much farm equipment is to be handled, a suitable ramp is one con-

Fig. 7.25—Multi-purpose sheep handling ramp. Notice the wide gate which swings forward in order to crowd sheep into the ramp.

structed of concrete and wide enough to handle heavy farm machinery. This width should be 10 to 12 feet. One side of the sheep loading fence should be built so it is removable, thus making possible the handling of equipment.

Build inside smooth—Not only should smooth lumber be used where sheep are to be crowded, but posts and braces should be on the outside to prevent injury to sheep. This, also, results in simpler, stronger construction. Nails must be driven well in and the use of carriage bolts with the heads on the inside is recommended. It is best if the chute is boarded solid, as it tends to encourage the movement of sheep and discourage their trying to get out.

Use shipping crates—Purebred breeders or those buying a single ram or so can utilize a shipping crate to good advantage to transport animals. The shipping crate shown in Fig. 7.26, when made of well-seasoned pine, weighs only 25 pounds.

Hay for feeding while in transit can be placed between the crate and a gunny sack tacked on the outside of the front end.

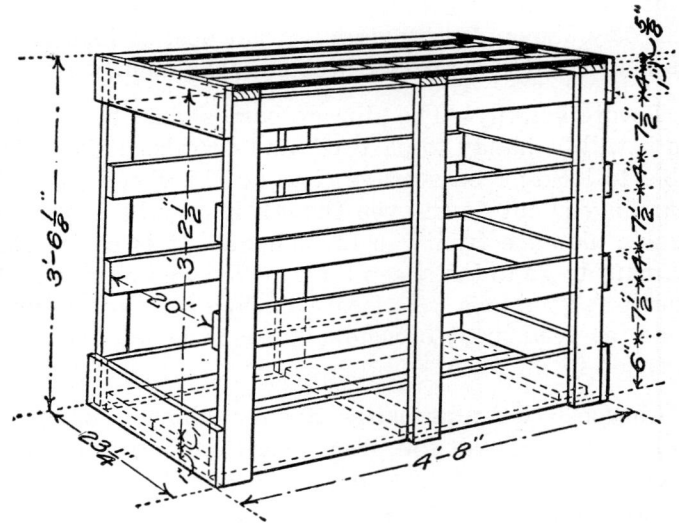

Fig. 7.26—Lightweight sheep crate. (Courtesy, USDA)

10. Providing Suitable Feed Storage

Feed storage is generally not as great a problem with sheep as with other kinds of livestock. This, of course, is due to their ability to forage most of the year and the fact that concentrates do not make up a high proportion of the diet for most flocks. On the other hand, economical storage, ease of handling feed in storage, and getting it to livestock, especially in adverse weather, are problems facing many producers. Adequate planning and the use of approved practices will aid greatly in overcoming this problem.

Keep construction simple—Kansas Bulletin 316 says that framework is the cheapest and most logical construction for sheep barns. Generally speaking, the simpler the construction, the more efficient it is. Loft construction has many advantages as it not only provides storage, but protection space underneath as well. The hay above also provides a great deal of insulation as heat is often a greater problem than cold.

Locate handily—The storage barn should be located so it will be handy to pastures and corrals. It is desirable to locate feed storage close to the sheep so that handling is cut down even though it may be a considerable distance from the farm home. It is easier for humans to go to the sheep than move large numbers of sheep. A good idea is to locate storage near an all-year road, as feed must occasionally be moved to another location in the event of an emergency.

Choose well-drained site—Shelter which is damp is unhealthy and unfit for sheep. For this reason, storage should be located in a well-drained area because in all probability sheep will be penned close to feed storage.

Locate on level areas—If necessary, it is wise to grade an area before starting construction. Don't guess; use a level or transit. Storage facilities built on unlevel ground tend to become lopsided as the ground washes away. Hillsides are not objectionable as building sites and may even prove an advantage from the standpoint of loading and unloading, providing the area where the storage barn actually sits is graded level.

Consider silage—In many areas silage is not an important feed for sheep, yet it is an excellent feed for them. **California Circular 411** says silage conserves

SHELTER AND EQUIPMENT FOR SHEEP

nutrients, provides emergency feed, and utilizes crop by-products. Silage has certain advantages over haymaking in that it conserves a greater proportion of nutrients and can be made during most weather conditions. Two to three pounds of silage daily per ewe will keep them in good shape.

Build silo to fit needs—Size depends upon number of stock to be fed daily. Graphs and tables aid in determining the size. Silos can be made of a great many different materials and styles. Tower silos, pit silos,

RELATION OF HERD SIZE TO DIAMETER OF SILO, BASED ON 40 POUNDS OF SILAGE PER CUBIC FOOT AND THE REMOVAL OF 3 INCHES OF SILAGE DAILY TO AVOID SPOILAGE

Inside diameter of silo	Volume per foot of depth	3 inches daily
(feet)	(cubic feet)	(pounds)
10	78.5	785
12	113.1	1,131
14	153.9	1,539
16	201.0	2,010
18	254.5	2,545
20	314.2	3,142

To find the number of animals that may be fed daily, divide by 2 or 3 (lbs.) depending on amount each animal is to have.

temporary silos such as snow fence, trench silos, etc., are all adequate for storing silage. Some operators have simply piled silage out on the ground and covered it with straw and dirt with fairly satisfactory results. Furthermore, silage can be made of almost any plant material including green grass. The disadvantage of silage is, of course, the labor required to put up and feed it plus the fact that it cannot be transported economically to different parts of the farm.

Use woven wire—Barbed wire tears the wool and will not prevent sheep from going through under most conditions. For this reason, woven wire must be used in order to make fence sheep-tight. Woven wire fencing, measuring 26″ to 32″, is recommended. Mesh should be small enough so that sheep cannot put their heads through and get caught. At least five strands of barbed wire must be used if a fence is to be sheep-tight and even this is not as satisfactory as woven wire. About two or three strands of barbed wire should be placed above the woven fencing in order to keep other domestic stock and dogs out. Another recommendation is to purchase higher woven wire fencing [39″ to 42″] and use only one strand of barbed wire on top. The additional cost of higher fencing will be compensated for by the reduced cost of barbed wire.

Build good fences—This means that first costs will be high; however, it is sound economy in the long run as labor costs are the same for either cheap or good fencing. Temporary or cheaply constructed fencing only encourages sheep to break out, and this is habit forming and in the long run such fences will have to be built over anyhow. Posts should be set 12 to 15 feet apart with proper corner posts and double-braced posts approximately every one-fourth mile. Posts more than 16′ apart generally require a 2x3 stake at 8′ from the posts. Good posts are set up to $2\frac{1}{2}$ feet into the ground although shallower depths are desirable under some soil conditions.

Treat posts—Every post should be treated with a preservative before setting into the ground—10 to 20 years of life will be added to the life of a fence post by so doing. Creosote or some of the newer products, like "penta," are satisfactory preservatives. Split posts last longer than sawn posts although possibly are not as attractive. Poles are the poorest type post although often the only kind available. Therefore, it is imperative that poles used for posts be well treated with a preservative. Redwood and cedar are superior to most timber used for posts although almost any kind will do. Split oak posts, while ragged looking, are excellent, especially when the fence line must be run through standing water.

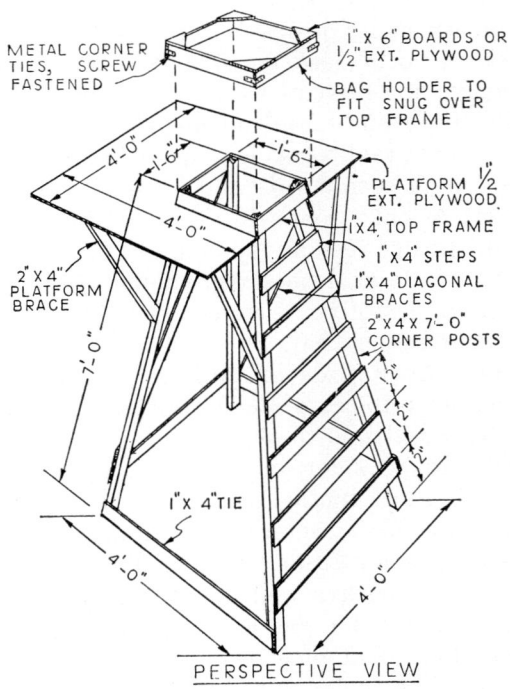

Fig. 7.27—A wool packing rack.

Consider steel posts—Fences built with steel posts are attractive, strong and long lasting. The cost for posts is often higher than wood, especially when many ranchers can cut and split their own posts. On the other hand, steel posts do not rot underground and therefore outlast wooden posts many times over. Fences do not sag as frequently with steel posts because the nails in wooden posts tend to loosen and permit the fence to shift. Other important advantages are that steel posts can be "driven" in almost all soils resulting in rapid fence building and can be used over and over if the fence has to be moved.

Build concentrate storage rodent-proof—Rodent-proofing is especially desirable for concentrate storage. Small grains, cottonseed meal, and other by-products are expensive and difficult to store economically unless kept out of the rain and in rodent-proof buildings. It is surprising how great a loss will occur in a year if small openings, which permit rats and mice to enter, are not sealed over. A leaky roof will often go unnoticed until large quantities of feed are spoiled. A further loss may occur in the form of dead sheep should spoiled feed be used. It is a desirable practice to build concentrate storage facilities so that they are bird-proof as well. Pigeons and other birds in addition to rats and mice have been known to carry and spread disease which can at times be disastrous to sheepmen.

Materials to consider in constructing rodent-proof facilities are sheet metal, especially for roofing, concrete, hardware cloth, screen, pipe, tongue and groove lumber, and exterior grade plywood. Tight joints, especially around doorways and openings, are a must as mice can easily enter a one-half inch opening. For this reason, one-half inch pipe or larger used for electric conduit should be sealed with wire screen or hardware cloth.

Utilize bin storage—For the small farm flock such storage is entirely adequate and oftentimes the most economical. Even for range sheepmen, it has a place in the storage of salt or limited amounts of cottonseed meal or other concentrates. Bins may be purchased commercially and, as such, are generally constructed of galvanized metal, or they may be home built. In the latter case, T & G lumber, exterior-grade plywood, or sheet metal may be used satisfactorily. Welded steel bins are excellent. It is important that all joints be tight, that it be well braced so as to stand moving about, and most of all, that a rain shedding, waterproof cover be part of the bin. Large, galvanized metal garbage cans are top-notch, but are limited in storage capacity and more expensive per cubic foot of storage space than other types. One advantage of bins is that they will work fairly well in the open although most sheepmen prefer to place them under an open shed. Commercial grain storage bins about 12 feet in diameter and 10-12 feet high are excellent for storing all sorts of concentrates, salt, and other feeds, although they are considered to be too expensive by many producers. If brick, basalt blocks, adobe blocks, or other such materials are available, ranchers can utilize their own labor to advantage in constructing rodent-proof facilities for grain storage. These buildings are very desirable and will pay out in the long run.

FEED SPACE REQUIRED

Self-feeders .... 1 inch per animal

Hay and grain.............................15 inches to 18 inches per ewe

Hay and grain.................................12 inches per lamb

Lambing pen.................................12 to 16 square feet per ewe

Space under shelter.........................10 to 15 square feet per ewe

Water should be available at all times.

CHAPTER VIII

CONTROLLING PARASITES AND DISEASES

Many times the difference between profit or loss in sheep production can be traced directly to a parasitic or a disease condition. Sheep are especially prone to parasitic infestations, and while animals may not die or produce obvious symptoms, the slow, constant drain on their health often means a substantial loss in income to the producer. People who show a profit year after year in the sheep business are generally those who also are aware of and make an effort to control parasites and diseases within their flocks. For these reasons, beginners in particular should see to it that many approved practices are carried out in regard to management of their livestock.

Activities Which Involve Approved Practices

1. Providing a satisfactory prevention and sanitation program.
2. Controlling dogs and predators.
3. Preventing and relieving bloat.
4. Controlling parasites.
    a. Internal
    b. External

5. Controlling shipping fever.
6. Controlling foot rot.
7. Controlling anthrax.
8. Controlling lamb dysentery.
9. Controlling blue bag (mastitis).
10. Controlling pink eye.
11. Controlling blackleg.
12. Controlling bloody scours (coccidiosis).
13. Controlling stiff lambs.
14. Preventing losses from poisonous plants.
15. Controlling scrapie.
16. Controlling bluetongue.
17. Controlling milk fever.

1. Providing a Satisfactory Prevention and Sanitation Program

Unlike humans, animals are difficult to diagnose and treat when ill. Because of their nature and the fact that they resist handling and treatment, many times a

Fig. 8.1—Well-grown, healthy animals such as the above on good pastures have a surprising amount of resistance to parasites and diseases. The best way to keep animals healthy is a sound prevention program.

sick animal is almost a dead animal. This is particularly true of sheep which are probably the most helpless of domestic livestock. In addition, it is hard even for conscientious breeders to keep them under close enough surveillance to treat them early. For these reasons, it is imperative that the prevention part of disease control receive the utmost consideration.

Maintain healthy livestock—Once an animal goes below par in weight and general resistance, it is easy prey to a host of parasites and infections. It is surprising how healthy, well-fed animals will successfully resist parasites and diseases under a wide range of conditions. Should these animals become ill, they often have a reserve to carry them through until they reach normal health again. It is not necessary or advisable that sheep, especially breeding stock, be overly fat. However, they should be active and alert with a reasonable amount of condition or flesh, be good eaters, and have regular bowel movements. Good pastures, supplemental feeding when necessary, properly trimmed feet, and available minerals will do much to maintain proper health.

Rotate pasture—This is good insurance against parasitic infestations. Usually it is a more satisfactory arrangement from the standpoint of pasture management as well. However, the main advantage is to prevent a build-up of internal parasites, as rotation tends to break the life cycle of these organisms.

Eliminate mechanical hazards—Old wire, glass, cans, broken equipment, etc., should be eliminated from wherever sheep can come into contact with such items. This is particularly true in corrals or chutes where the sheep are apt to be crowded and worked. Any cut or bruise is a possible source of infection that can lead to fatal results. Bogs, deep holes, old wells, or dangerous cliffs should be fenced off so that sheep cannot crowd around and push one another in, as frequently happens around unguarded hazards.

Observe them continually—Unlike other farm animals, sheep are completely helpless, oftentimes in ridiculous situations. Fat sheep will lie down in irrigation ditches and suffocate because they cannot turn over to get out. Hoofs should be trimmed regularly, as animals always on soft, marshy ground grow hoofs too long and their feet become too sore to walk on. Eyes, nostrils, and the rectal area should be watched and inspected to see that wool maggots, screw worms and other parasites are not present.

Prevent wool blindness—Breeds with wool on their faces may be unable to see properly when the wool

Fig. 8.2—A wool blind sheep. It is a desirable practice to trim the wool away from the eyes as a wool blind sheep cannot graze properly and thus becomes thin and weak.

becomes too long. Whenever this happens, they refuse to graze or cannot find enough food thus becoming thin and weak. It is a desirable practice to trim this wool away periodically so normal grazing habits are possible.

Practice sanitation—Good, general, sanitary measures should always be utilized in order to prevent widespread epidemics among the flock. Dead sheep should be buried so that dogs or carrion feeders cannot spread infection. Some diseases like anthrax are dangerous to humans, and in such cases the pelts should not be removed but the entire carcass burnt and buried deep. Corrals, lambing sheds, and similar areas should be opened to the sun and dried as much as possible. If occasion warrants, the area should be thoroughly disinfected to kill all germs. This process is frequently not done properly, and false confidence is placed in it, as proper disinfecting of quarters is a difficult thing to do. All instruments used in treating sheep should be well sterilized before using on other animals. It is not a good practice to borrow equipment from a neighbor. If disease is a problem, visitors should be discouraged from entering sheep premises until normal health in the district is restored.

2. Controlling Dogs and Predators

Many persons considering entering the sheep business would never consider this factor of any importance. This is a natural conclusion. However, both are current dangers and very often under surprising situations. Family pets, friendly dogs, and otherwise well-mannered canines frequently turn out to be sheep-killing dogs. Unsuspecting owners of such pets can hardly believe such is the case until presented with proof. Unfortunately, once a dog or pack of dogs start killing sheep, it is practically impossible to break them of the habit.

Fig. 8.3—Predators trapped by a government trapper. Coyotes and bobcats can inflict serious economic loss to sheepmen. Frequently a lone marauding animal will be responsible for most of the damage on a single ranch.

Natural predators, including coyotes, cougars, bobcats, wolves, and the like, have long been known to stockmen for their sheep-killing tactics. Laws, bounties, government trappers, and hunters have been employed to keep the number down with varying degrees of success. In spite of this, each year there is a substantial economic loss incurred by depredations from these marauders. While most of the native wildlife on the continent has restricted its range, some of the predators, especially the coyote, have enlarged their range and habitat compared to primitive times. In fact, it is revealing to find out how well these animals can get along in civilization, frequently in the shadow of a large city.

For these reasons, utmost precaution must be utilized to prevent economic loss by predators.

CONTROLLING PARASITES AND DISEASES

Consult your state law—Almost every state has some regulations on its statutes relative to pets and dogs, especially where sheep raising is an important enterprise. Knowledge of these laws can guide people in controlling their own pets as well as helping them to realize their rights in regard to protecting their flocks from other pets and preventing damage.

Build dog-proof and wolf-proof fences—It would be too expensive to enclose large tracts of land with such fencing. However, small enclosures made dog- and wolf-proof can be constructed and the flock can be driven into it each night should the danger warrant it. Obviously, this method has many limitations. Nevertheless, there are times and areas where it is the only way. **Farmers' Bulletin 1268** gives excellent plans for building such a fence.

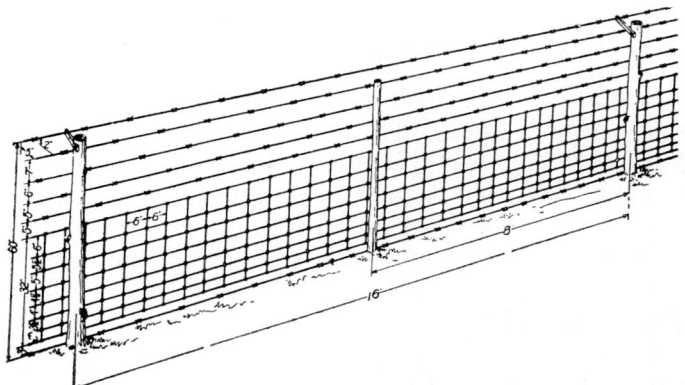

Fig. 8.4—A dog-proof and wolf-proof fence. (Courtesy, USDA)

Eliminate sheep-killing dogs—This does not always mean destroying the animals, as they may be perfectly reliable and safe in every other way. However, they must at least be removed to an area where there are no sheep, or else completely confined. The most important information to remember is that sheep-killing dogs will teach other dogs to do the same thing; therefore, most practical sheepmen advocate destroying such animals in a humane manner.

Limit number of dogs—There is a tendency for many homes to have too many dogs. Therefore, it is desirable to keep this number within reasonable limits and thereby lower the potential number of sheep-killing incidents.

Consult government hunters—These individuals are generally only too glad to respond to calls whenever natural predators get out of hand. It is amazing, at times, to see the number of predators they can trap in a given area. In addition, they can often give valuable advice on how the rancher can trap or hunt and control these predators himself. The State Division of Fish and Game will direct ranchers to the proper persons to consult.

3. Preventing and Relieving Bloat

Bloat is a common disease of sheep. In a sense it is more acute in sheep than in cattle because of a lesser chest cavity, thereby causing death more quickly than a similar condition in cattle. This condition is the result of an abundance of gas formed by food fermentation in the paunch. In such cases, it is formed more rapidly than it can be expelled by belching, with a resultant bloated condition.

Range sheep or sheep feeding on dry pastures are not troubled to any great extent with bloat except, perhaps, during the lush growth period in the spring time.

The difficulty with sheep is that with such large numbers, affected animals are often overlooked until it is too late. For this reason, the prevention angle must be heavily stressed. There is a great variation in individual animals in their susceptibility towards bloat.

Provide non-bloating pastures—This is one of the

most practical answers to bloat prevention. Pastures that contain at least 50% by weight of grass in a grass legume pasture give good results with bloat prevention. In some cases, sheep will prefer the grass clover pasture to straight clover if given a choice.

Don't change rations rapidly—Whenever animals have been on one kind of ration and are suddenly changed to another, bloat is likely to occur. This is especially true when changing over to a succulent ration containing large amounts of green feed. Once sheep have been accustomed to grazing green pastures, the danger is lessened as they seem to become accustomed to the feed, and feed continually on small amounts so that bloat is greatly curtailed. Therefore, it is desirable to change over from one feed to another slowly and be constantly alert should trouble develop.

It is also advisable to leave animals continually on rich pasture once they are used to it rather than keeping them off for a certain period of the day. This is especially true if animals tend to get hungry when kept away from pastures.

Guard against spoiled feed—This often upsets the stomach and general health of the animal so that bloat is more likely. Occasionally spoiled feed alone will cause death.

Guard on lush feeds—Alfalfa, clovers, and other legumes, even green grass, beet tops, etc., are the prime producers of a bloated condition. Fields wet with dew or feed bunched and undergoing heating seem more likely to be dangerous, although there is no scientific proof of this.

Provide some coarse feed—Stimulation of the stomach lining by fibrous material starts belching and prevents bloat. For this reason, whenever bloat is a danger, there should be access to coarse feed. Dry hay, stubble, cured stemmy alfalfa, or other coarse feeds will prevent bloat to some extent. This is recommended because animals seem to crave some dry feed whenever grazing succulent pastures entirely.

Provide proper diet—An adequate supply of minerals plus a balanced diet helps eliminate bloat danger, because animals frequently eat an oversupply of other available feed whenever a shortage in one area exists. Salt in particular should be provided as well as an ample supply of water and minerals.

Avoid change and excitement—Digestion may be upset by emotional changes which often result in bloat. Therefore, it is desirable to maintain routine, calm conditions and refrain from causing any excitement.

Use a gag—This consists of a stick tied in the animal's mouth to induce swallowing, salivation, and increase belching. It is not used with sheep as much as with cattle. Nevertheless, valuable animals oftentimes may be saved by this practice.

Drench the animal with surface tension reducers—A pop bottle of kerosene and milk is a good farm remedy, as it reduces surface tension of gas bubbles and stops fermentation for a time. Many other substances can be used that veterinarians will advise and which can be obtained commercially. However, two ounces of turpentine in a quart of milk is a standard remedy.

Use a knife as the last resort—This is good advice, but once the animal is in critical condition, do not hesitate to "stick" it, as it will probably die shortly if you do not do so. Worry about abscesses later. A veterinarian could do a better job, but time is essential. Therefore, puncture the left side of the animal with a thin-bladed butcher knife in the center of the triangle formed by the hip bone, last rib, and the back muscle.

CONTROLLING PARASITES AND DISEASES

Give the knife a twist to let gas escape, but don't let it escape too rapidly.

Keep animals moving—Driving animals up hill or refusing to let them lie down aids in preventing death loss from acute bloat. Occasionally a stream of water can be sprayed on the side of the animal or the entire herd driven into a stream. This stimulates the stock and keeps them active, thereby inducing belching.

4. Controlling Parasites

Parasites probably account for a greater economic loss with sheep than any other kind of livestock. First, there are generally large numbers of animals involved so that trouble goes unnoticed; second, due to their close grazing habits they tend to pick up worm eggs easily; and third, their banding habits keep them together so that infestation spreads rapidly. Approved practices in controlling parasites will pay off well with sheep. There are two main types of parasites affecting sheep:

a. Internal—parasites found inside the body.
b. External—parasites found on the outside of the animal on the skin or in the wool.

Controlling Internal Parasites

Since the advent of permanent and irrigated pastures, internal parasites have become an increased danger because of moist conditions ever prevalent. While there are a large number of parasitic species involved, the control and treatment is rather similar. For this reason, they will be considered as a group.

Internal Parasites

Practically all farm flocks of sheep have some inter-

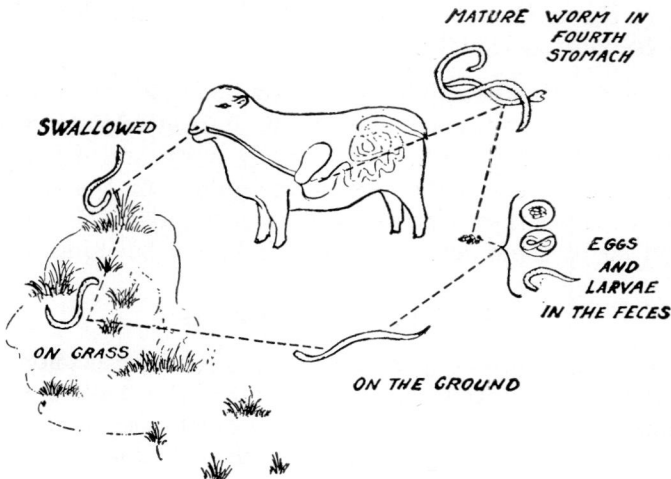

Fig. 8.5—Life cycle of the common stomach worm (Haemonchus). The mature stomach worm lives in the fourth stomach (abomasum). The female lays eggs which are discharged in the feces. If the feces and eggs are dropped where there is suitable moisture, a larva develops in the egg which later hatches and undergoes several moults on or in the ground. It is then ready again to infect sheep. With moisture from rain or dew on the grass, the infective larvae move up on the grass, where they enhance the possibility of being ingested by grazing sheep.

nal parasites. Internal parasites are a constant drain, and as a result, growing animals are unable to make profitable weight gains and, because of a run-down condition, are more susceptible to disease. Stunting and unthriftiness, and even death are resultant from heavy infestation. Bottle jaw or intense swelling under the chin is a common symptom of heavily parasitized sheep.

Poor doers, thin, emaciated animals, as well as all sheep from known areas where worms are a problem should be suspected of being infested and handled accordingly.

Break life cycle of parasites—Good management and

CONTROLLING PARASITES AND DISEASES 257

proper sanitary measures can prevent and control constant outbreaks of parasitism, as animals become

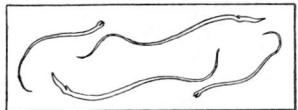

Fig. 8.6—Stomach worms look like coarse white hairs about one inch long. They suck blood from the lining of the stomach.

infested by eating feed, pasture, or drinking water containing worm eggs or larvae. Clean pens, houses, sheds, and stables will reduce the danger. Well-drained lots and pastures, clean drinking water, raised feed and water troughs, pasture rotation, and avoidance of overstocking all reduce the chances of parasites continuing to live long enough to reproduce.

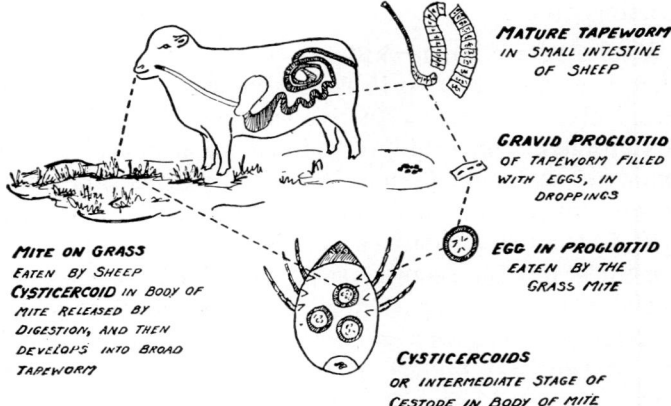

Fig. 8.7—Life cycle of the broad tapeworm of sheep. The mature tapeworm lives in the intestine of sheep. Segments are continually being broken off the posterior end and leave the animal with the feces. The proglottids are filled with eggs which when eaten by certain grass mites (Orbatid) develop as a cysticercoid in the body of the mite. Orbatid mites crawl up grass blades, where they are eaten by grazing sheep. In this way, sheep become infested or reinfested, and the life cycle of the tapeworm is completed.

Keep animals in proper nutrition—It is not necessary that they be fat, but good growing conditions strengthen the animals' resistance and give them a better chance to resist infestation. Creep feeding of lambs and supplemental feeding of the ewe flock may become necessary at times.

Treat all the herd—Since any parasites remaining in the herd will reinfest other animals, all of the flock should be treated twice a year. Early in the fall and in early spring are the best times. Lambs over three months old should be included.

Remove to clean pasture—Treated animals should, if possible, be put into clean pastures to prevent reinfestation. Crop rotation or exposure to bright sunlight is a good way to establish a clean pasture unless too much moisture is present. Therefore, it is desirable to coordinate treatment just before moving the flock to new grazing areas.

Use modern worm killers—Many new compounds have become available in recent years. Thibenzole and tramisol are two new effective wormers. They have the advantage of controlling a wide range of parasites. For example, stomach worms, intestinal worms, and lungworms can be controlled at one dosage. The disadvantage is that treatment costs slightly more than older remedies, but the results are so superior that the extra initial cost is very worthwhile. One of the first of the modern wormers was phenothiazine. Phenothiazine is one of the best known of the modern drugs used to combat internal parasites. It can be mixed into the feed or fed with salt. Occasionally, it is given in individual tablets or in a drench. If parasitism becomes a real problem, a veterinarian should be called for an accurate diagnosis and a more specific treatment, as flukes, different types of worms, and the like respond to some drugs better than others. It is a handy, safe, desir-

CONTROLLING PARASITES AND DISEASES

Fig. 8.8—Using a drench gun to drench a mature sheep. Thibenzole and other modern internal parasite compounds have greatly reduced the worm problem in sheep with resultant increase in profits. (Courtesy, Merck & Co., Inc.)

able practice to follow printed dosage and treatment recommendations of the manufacturer of the product. More than one treatment may be necessary in some instances. Sheep should not be kept off feed before treating with phenothiazine.

Liver fluke, according to **Texas A. & M. Circular 287,** is best controlled by one to two ounces of a suspension of hexachlorethane prepared by thoroughly mixing:

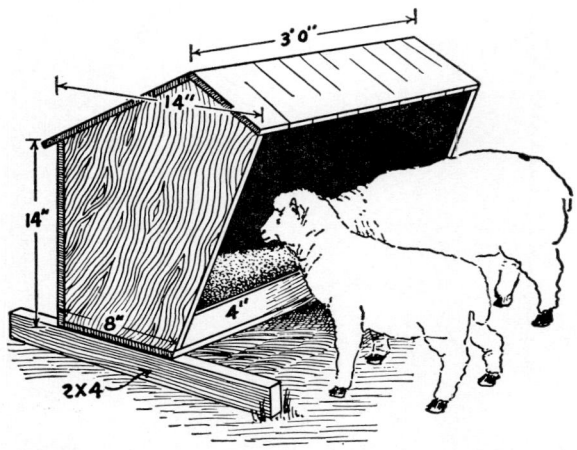

Fig. 8.9—This covered salt box is ideal for feeding the salt-phenothiazine mixture to ewes and lambs on pasture.

> Hexarchlorethane .... 1 pound
> Bentonite .... 1.5 ounces
> Water .... 1.5 pints

One dose is usually sufficient given individually with a dose syringe.

Drench small flock—It pays to drench even large flocks individually, and common practice is to drench small flocks when conditions demand. **Oklahoma Circular OP-39** states that phenothiazine will not control tapeworms and recommends the following treatment: Dissolve one ounce of bluestone in one quart of hot water, and one ounce of snuff (or Blackleaf 40) in one quart of tap water. When ready to drench, mix the two solutions and add one quart of cold water. This makes enough for 20 to 24 adult sheep, or 45 to 48 lambs. (Note: Only glass or crockery should be used, as bluestone corrodes metal.)

Drench repeatedly for tapeworms—Keep sheep off feed 18 to 24 hours before drenching, and two to four

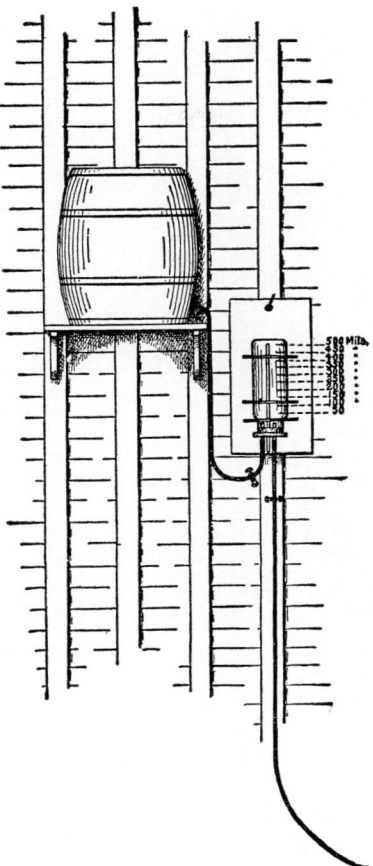

Fig. 8.10—Dosing device for administering copper-sulfate solution.

hours afterward. Drench every 28 days, after grass begins to grow in the spring, and repeat until first hard freeze in the fall, if tapeworms are a serious problem.

Provide phenothiazine—salt lick—Texas Circular 287 says that a mix composed of 90 pounds salt and 10 pounds phenothiazine supplied free choice to sheep and

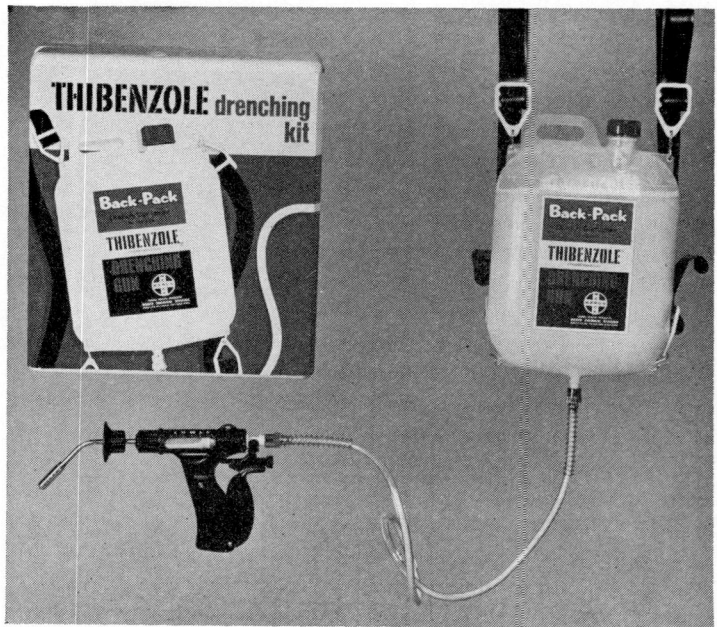

Fig. 8.11—Large numbers of sheep can accurately be dosed in a minimum amount of time by the use of a drenching gun system like the one shown. (Courtesy, Merck & Co., Inc.)

goats results in a fair measure of control, as 95% to 100% of the roundworm eggs passing do not hatch when this treatment is used. It is important to provide fresh phenothiazine every two weeks as it rapidly changes into other compounds upon exposure.

External Parasites

Included under this category are ticks, lice, screw worms, scab mites, and blow flies. All authorities agree that these pests are an important factor in influencing the production of sheep, lamb, and wool. Symptoms are somewhat similar in the appearance of animals to that of internal parasitism except that the parasites are

PARASITE CONTROL ON SHEEP

PEST	When To Treat	Chemical	Formu-lation	Concen-tration	Amount	Method of Application	REMARKS
SHEEP KED	Spring, following shearing	Coumaphos (Co-Ral®)**	WP	25%	8 lb/100 gal water (0.25%)	Spray	Do not treat animals under 3 months old. Lightly spray animals 3-6 months old. Do not use with pyrethrins, allethrin, or synergist. Do not spray animals for 10 days before or after shipping, weaning, or other exposure to disease. Do not apply within 14 days of phenothiazine treatment. Do not market animals in less than 15 days after treatment.
		Dioxathion (Delnav®)	EC	30% (2 lb/gal)	½ gal/100 gal water (0.15%)	Dip or spray	Do not apply more than once every 2 weeks. Do not dip animals under 3 months old.
		Lindane**	WP	25% gamma isomer	0.8 lb/100 gal water (0.025%)	Dip	Do not market animals within 1 month after spray treatment or 2 months following dipping. Do not dip animals under 3 months. Repeat at 2- to 3-week intervals if needed.
			EC	25%	1.6 lb/100 gal water (0.05%)	Spray 4-6 qt/animal	
		Malathion	EC	55% (5 lb/gal)	7 pt/100 gal water (0.5%)	Spray thoroughly	Do not use on animals less than 1 month old or on sick or lactating animals.
			WP	25%	16 lb/100 gal water (0.5%)		
		Ronnel** (Korlan®)	EC	25% (2 lb/gal)	2 gal/100 gal water (0.5%)	Spray	Do not market animals within 3 months of treatment. Do not reapply within 2 weeks.
		Toxaphene**	EC*	72% (8 lb/gal)	½ gal/100 gal water (0.5%)	Dip or spray	Do not market animals within 1 month of treatment.

* The 72% (8 lb/gal) emulsifiable concentrate is recommended because of lower cost and greater stability in final mix form.

** Repeat at 2- to 3-week intervals if needed.

APPROVED PRACTICES IN SHEEP PRODUCTION

PEST	When To Treat	Chemical	Formu-lation	Concen-tration	Amount	Method of Application	REMARKS
WOOL MAGGOTS (Fleece worms)	Warm, humid weather in areas having history of wool maggot problem	Coumaphos (Co-Ral®)	WP	25%	8 lb/100 gal water (0.25%)	Spray	Safety precautions as given for sheep ked (sheep tick).
		Dioxathion (Delnav®)	EC	30% (2 lb/gal)	½ gal/100 gal water (0.15%)	Dip or spray	Do not apply more than once every 2 weeks. Do not dip animals less than 3 months old.
		Lindane	EQ 335 smear	3%	1 part to 9 parts water	Wet infested area and 3" around it.	Do not treat sick or lactating animals.
TICKS (Wood ticks)	As needed	Coumaphos ** (Co-Ral®)	WP	25%	8 lb/100 gal water (0.25%)	Spray	Safety precautions as given for sheep ked (sheep tick).
		Lindane	WP	25%	0.8 lb/100 gal water (0.025%)	Spray or dip	Do not treat sick or lactating animals. Do not market in less than 1 month after treatment.
		Toxaphene	EC *	72% (8 lb/gal)	½ gal/100 gal water (0.5%)	Spray or dip	Do not market in less than 1 month after treatment.
SPINOSE EAR TICK	As needed	Coumaphos (Co-Ral®)	Dust	5%	Inject into ears from squeeze-bottle duster as prepared by manufacturer.		
		Dichlorvos (DDVP)	EC	20%–24% (2 lb/gal)	½ oz/2 qt mineral oil (0.25%)	Oilcan	Apply ½ oz/ear by means of spring bottom oilcan with rubber-tipped spout.
SCAB MITES	When observed	Lime sulfur	Proprietary solutions			Vat dip for 2-3 minutes	Temperature of dip to be held at 95 to 105 F.
		Nicotine sulfate	Dilute according to instructions on container.				Temperature of dip to be held at 95 to 100 F.
		Toxaphene	EC *	72% (8 lb/gal)	½ gal/100 gal water (0.5%)	Dip not less than 1 minute.	Same as for beef cattle.
LICE	In fall or after spring shearing	Coumaphos (Co-Ral®)	WP	25%	8 lb/100 gal water (0.25%)	Spray	Do not market animals within 15 days of treatment. Repeat at 2- to 3-week intervals if needed.
		Lindane	WP	25%	0.8 lb/100 gal water (0.025%)	Dip	Do not dip animals less than 3 months old. Do not market animals within 1 month after spray treatment or 2 months following dipping. Do not use on sick or lactating animals.
					1.6 lb/100 gal water (0.05%)	Spray	

* The 72% (8 lb/gal) emulsifiable concentrate is recommended because of lower cost and greater stability in final mix form.

** Repeat at 2- to 3-week intervals if needed.

CONTROLLING PARASITES AND DISEASES

LICE continued	In fall or after spring shearing	Methoxychlor**	WP	50%	4 lb/100 gal water (0.25%)	Dip	No time limitation.
					8 lb/100 gal water (0.5%)	Spray	
		Ronnel (Korlan ®)	EC	24% (2 lb/gal)	1 gal/100 gal water (0.25%)	Spray	Do not market animals in less than 3 months after treatment.
		Toxaphene**	EC*	72% (8 lb/gal)	½ gal/100 gal water (0.5%)	Dip or spray	Do not market animals within 1 month of treatment.
SHEEP BOT FLY (nasal bot)	No effective treatment.						
SCREW-WORMS	Spring, summer and early fall	Coumaphos (Co-Ral ®)	WP	25%	8 lb/100 gal water (0.25%)	Spray as needed to wet animal.	Safety precautions as given for sheep ked (sheep tick).
			Dust	5%	Direct application	Dust wound thoroughly.	No time limitation.
			Spray	3%		Pressurized spray to wounds and surrounding area.	Do not treat within 14 days of slaughter.
		Ronnel (Korlan ®)	EC	24% (2 lb/gal)	2 gal/100 gal water (0.5%)	Spray as needed to wet animal.	Do not market animals in less than 3 months after treatment.

*The 72% (8 lb/gal) emulsifiable concentrate is recommended because of lower cost and greater stability in final mix form.

**Repeat at 2- to 3-week intervals if needed.

ABBREVIATIONS USED IN TABLES

gm - - - - - - - - - - - - - - - - - gram(s)
cc - - - - - - - - cubic centimeter(s)
oz - - - - - - - - - - - - - - - - - - ounce(s)
pt - pint(s)
qt - quart(s)

gal - - - - - - - - - - - - - - - - - gallon(s)
lb - - - - - - - - - - - - - - - - - - - pound(s)
tsp - - - - - - - - - - - - - - - teaspoon(s)
tbsp - - - - - - - - - - - - - tablespoon(s)
/ - per

EC - - - - emulsifiable concentrate
WP - - - - - - - - - - wettable powder
sol - - - - - - - - - - - - - - - - solution
psi - - - - - pounds per square inch
cwt - - - - - - - - 100-lb body weight

(Courtesy, Control of External Parasites of Livestock, 1970, University of California Agricultural Extension Service)

more easily observed and their depredations more quickly noticed. Dry, mangy condition of the wool as well as patches of it falling out is often encountered in heavily infested flocks. Approved practices will greatly reduce the losses from these parasites.

Prevent wounds—Blow flies and screw worms are particularly attracted by the smell from open wounds and lay their eggs under such conditions. Therefore, it is imperative to do everything possible to prevent breaks in the skin. Old wire, protruding nails, broken machinery, poor gates, and the like should be repaired or eliminated. Corrals and chutes into which animals are crowded and worked should be examined and made as smooth as possible on the inside to prevent injury or tearing of the skin.

Handle during the right time of year—Docking, castration, shearing, etc., should be done during the cool seasons. Early spring or late fall, before there is danger of fly strike, is best. Wherever there are burrs or foxtails, as well as external parasites to contend with, it is advisable to shear twice a year.

Tag to keep clean—Proper tagging around the vent and rear end as well as around eyes and nose in some breeds prevents the accumulation of dung and body secretions which give off odors attracting flies. If these areas are clean, sheep stay healthier and external parasites are cut down because of a lack of favorable areas in which to grow and lay their eggs.

Prevent fly strike—Screw worm smear should be painted on all open cuts or wounds whether such abrasions are the result of cuts from castration, other operations, or accidental wounds. This precaution is imperative in warm weather although it may be eliminated in cold climates or in the fly-free season. However, animals should be observed frequently, as fly strike

CONTROLLING PARASITES AND DISEASES

Fig. 8.12—These sheep have just been shorn and all have a number of small nicks and cuts. They must be carefully watched and should have a fly strike preventative applied if during strike season.

often occurs when least expected. Commercial preparations are available for this purpose. The United States Department of Agriculture has done considerable investigation of this problem and has developed a number of effective preventatives. One recently developed smear known as screw worm remedy E.Q. 335 is obtainable at veterinary supply houses. This smear is a combination of 3% lindane and 35% pine oil. According to **Texas A. & M. College L-131,** this material should be used sparingly and not more than one tablespoon used on the average size wound.

Control fleece worms—Commonly known as wool maggots, they are the larvae of several species of blow flies. They infest urine-soaked wool around the rump, and may cause subsequent infestation by screw worms.

Fig. 8.13—A dipping vat used to control external parasites. Sheep are forced to jump in and climb out so that the entire animal is treated. Every animal in the flock should be treated. (Courtesy, A. L. Mathis, Teacher, Vocational Agriculture Department, Kearny, Nebraska)

Control these by clipping away most of the infested wool and treat with screw worm remedy E.Q. 335, diluted one part to nine parts of water.

Control sheep ticks—This pest is really a wingless fly rather than a true tick. However, it is very common in all parts of the country. Injury to the wool and general health is serious. Lambs may even die as a result.

Control scab mites—Sheep scab is caused by a tiny scab mite. According to **Kansas Circular 212**, the symptoms are intense itching with loss of wool and formation of scabby areas under the wool and in denuded areas. It spreads rapidly from animal to animal. The United States Department of Agriculture recommends that each ewe swim for two minutes and have her head immersed twice. Use five pounds of BHC, 10% gamma isomer wettable powder to 100 gallons of water as a recommended mix. Water should not be warmer than 85 degrees for best results.

5. Controlling Shipping Fever

This is an infectious disease which often terminates in death. It is known medically as hemorrhagic septicemia because it results in a poisoning of the blood.

Cause—Generally, it is associated with hardships and the hazards of shipping. Most authorities believe that a group of microorganisms or a virus are the actual causative agents that invade the bodies of animals when their resistance is low as a result of hardship.

Symptoms—High temperature, loss of appetite, depression, discharge from eyes and nose, rapid, labored breathing, gaunt appearance, coughing, and sometimes diarrhea. Animals may die suddenly or develop pneumonia. Death loss is often high and survivors recover slowly. Prompt examination by a veterinarian may save many animals. The fever usually occurs after sheep have been subjected to hardship such as long drives or after shipment.

Prevention—Most effective prevention is elimination of conditions that tend to lower resistance. Overdriving, overcrowding, overfeeding, lack of rest, water, feed, and shelter during transit should be guarded against. However, it is possible to contact this disease if poor conditions exist on the home ranch even when transportation is not involved. Such factors as damp, unclean, drafty quarters will also foster the disease.

The use of biological products to prevent and treat outbreaks is still somewhat controversial. However, many successful livestock men and veterinarians believe these products to be a real aid. Prior to a long, hard trip, a great number of livestock men will vaccinate about a month before shipping, although with large numbers of sheep involved, proper handling is a much more economical precaution. Very valuable animals might well be vaccinated as an extra precaution, and a veterinarian called upon the slightest danger signal after prolonged hardship.

6. Controlling Foot Rot

This condition is a potential hazard wherever sheep are found, but is particularly prevalent during the wet season or in dirty, damp feed lots or swampy ground. Foot rot seldom causes death, but affected animals lose weight and may become too lame to get around. If, for example, such occurs to one or more of the rams during the breeding season, it could result in a greatly reduced lamb crop. Black faced breeds do not seem to be as prone to foot rot as others, but all are susceptible.

Cause—The malady is an infectious disease of the feet caused by microorganisms.

Symptoms—The first symptom observed is generally lameness ranging from slight to the point where animals refuse to stand. Later, swelling and soreness may be observed just above or between the claws or bulbs of the heel. A watery fluid with a characteristic foul odor may ooze from the swollen area. The foot may even rot off. However, high fever and animals off feed would be observed long before this occurs.

Prevention

Trim feet properly—This prevents breaks in the hoof and skin where organisms could gain entrance, as well as keeping the feet in good physical health.

Drain wet areas—Removing sharp stones, old boards with nails protruding, etc., will eliminate the hazards to a great extent. Muddy areas, swamps, and similar places, which harbor germs and keep the hoofs too soft so they can easily be bruised, should be drained or avoided.

Isolate new animals—New sheep added to the flock should be isolated for at least a month if foot rot is prevalent. It is advocated that replacements be obtained from flocks known to be disease free.

Place on dry ground—It is common practice to experience little trouble with foot rot on the range or on grain stubble, compared to the amount on animals grazed on irrigated pastures. For this reason, infected animals should be placed on dry ground immediately.

Treat infected animals—Although some animals recover spontaneously, the use of new, modern, germ killers is very effective. The sulfa drugs, prepared by various manufacturers, shorten the period of illness and generally speed recovery. Follow manufacturer's directions. Badly infected animals should have the diseased tissue removed and an antiseptic powder applied.

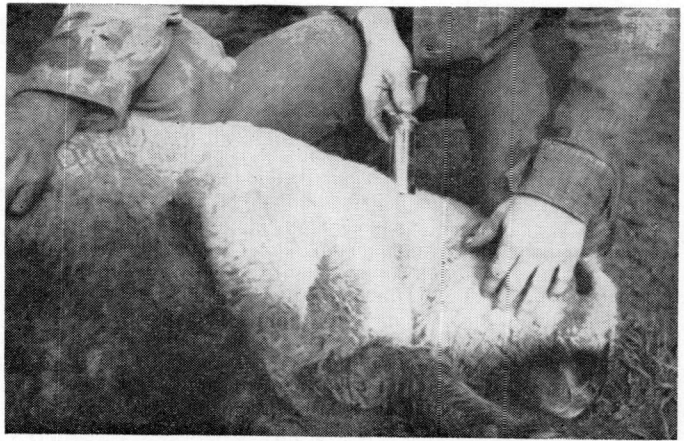

Fig. 8.14—Vaccination is an excellent means of prevention for many diseases. It should be done by, or at least under the supervision of, a competent veterinarian.

After the dead and diseased tissue is removed many sheepmen have found that a salve made of the following ingredients works quite well: vaseline (petroleum jelly), bluestone powder (copper sulfate), and sheep dip. The mixture should be smeared on and worked as deeply as possible into the affected area.

7. Controlling Anthrax

In its most common form this disease is characterized by a rapid, fatal course or sudden death. Unlike many livestock diseases, it is doubly serious because it is transmissible to humans. In this form it is known as wool-sorter's disease.

Cause—Anthrax is an infectious disease caused by a specific microorganism known as the anthrax bacillus. According to **Farmers' Bulletin 1736,** the germ is indigenous to the soil in certain areas where it survives for long periods, especially in low lying, marshy land. These

CONTROLLING PARASITES AND DISEASES

areas are known as anthrax districts. Outbreaks generally occur in late spring or summer, which is often known locally as the anthrax season. Infection comes by drinking contaminated water or eating food produced in an infected area. Flies, scavengers, and other incidental factors may spread the disease.

Symptoms—Visible symptoms are unsteady gait, trembling, difficult respiration, restlessness, bloody discharge from natural openings, and blueness of mucous membranes. Generally, the animal is found dead or dies shortly after symptoms are noted with convulsions preceding death. It may be confused with blackleg.

Prevention

Consult local authorities—This disease is serious enough so as not to be trifled with, and its identity should be established immediately. Veterinarians or local health authorities should be consulted once the disease strikes. In many places this is required by law.

Cautiously dispose of carcasses—Exercise care in handling infected carcasses. However, all dead animals, bedding, or other contaminated materials should be burned or buried deeply in quick-lime.

Vaccinate early—Just before the local anthrax season, animals should be immunized with anthrax vaccine. A single injection gives adequate protection for a year. In the early stages of the disease, some commercial preparation of penicillin may help sick animals. Vaccination should be under the supervision of a veterinarian.

Exclude danger areas—Fencing of pools, swamps, or pastures may be desirable where the disease is known to exist permanently.

Move to new pasture—Removing stock, such as large

range herds, to the mountains for summer grazing before the anthrax season may eliminate the need to vaccinate and reduce the danger of an outbreak.

8. Controlling Lamb Dysentery

This is a serious disease of newborn lambs. While it may occur in older animals, it generally develops when the lambs are one to six days old. Approved practices must be employed constantly, as it spreads quickly.

Cause—Lamb dysentery is caused by bacteria that are normally present in the intestinal tract of mature sheep. The infection is acquired by the lamb swallowing tags of manure from the legs of ewes as it nurses, or from soiled bedding.

Symptoms—Death may occur before symptoms are noticed. However, foul-smelling, whitish or gray-yellowish, blood-streaked droppings are common symptoms. Lambs fail to nurse and lose weight.

Prevention—It is a good practice to consult a veterinarian immediately in order to prevent heavy losses to the entire flock. The key preventive factor is proper sanitary measures. For this reason, range lambing (when climate permits) is an excellent method to help prevent the disease. Sick lambs and ewes should be separated immediately to clean quarters. Proper medication by individual treatment, with a compound such as sulmet, usually stops scouring and prevents further loss. One to four treatments at 24 hour intervals may be necessary depending on the severity of the attack.

9. Controlling Blue Bag (Mastitis)

This is a chronic condition of the udder whereby it is inflamed and swollen. It usually follows injury or mishandling of the mammary system.

CONTROLLING PARASITES AND DISEASES

Cause—One or several kinds of bacteria enter the udder through the teats or through cuts, scratches, or bruises.

Symptoms—Two forms may be exhibited. One is known as blue bag, which develops very rapidly and is very serious. It is characterized by a blue-violet color of the greatly swollen udder. This condition may carry over onto the legs and abdomen. However, the udder itself may not be painful to the ewe. Gangrene may develop and many animals die.

In the chronic or less acute form, the udder is swollen and painful with hardened areas. The milk in either case is watery or yellowish, and contains flakes of blood or pus-like material. Generally, the ewes refuse to eat and drink, walk with a straddling gait, and do not let the lambs nurse.

Treatment—The first precaution is to remove all possible hazards which can lead to cutting or bruising udders. Old wire, boards with protruding nails, improperly constructed jails and sheds should be eliminated so that udders are safe from mechanical injury. Cleanliness and good management, including clean, well-drained lambing pens reduce the incidence. According to **University of California Circular 130**, treatment should start immediately. Separate affected ewes from the rest of the band in order to inhibit spread of the infection. Sulfanilamide given orally, thirty grains in tablets three times daily, is recommended and reasonable results can be expected.

10. Controlling Pink Eye

This is a relatively common malady with sheep, but not particularly serious if adequate precautions and treatment are used. On the range, animals will lose considerable weight because of inability to find feed, and in any case, if untreated pink eye may lead to blindness.

Cause—University of California Circular 130 states that the cause may be divided into two groups: (1) An infectious microorganism, particularly streptococci and staphylococci, which may spread rapidly through the band, and (2) irritants, such as foxtails, thorns, dust, grass awns, that produce a similar condition. It is also true that faulty nutrition (vitamin deficiency), strong light, etc., may predispose animals to infection.

Symptoms—These are easily recognized. Eyes swollen, inflamed, protruding, and with a watery discharge are characteristic symptoms. Later, they may become cloudy, while eyelids tend to close and discharge pus. One or both eyes may be infected. Loss of appetite and fever are general reactions as well.

Prevention—Good nutrition plus removing animals from dusty, windswept pens will relieve some of the predisposing causes.

Segregate animals—If only one or two are affected, they should be separated to a quiet, darkened shed. Healthy range sheep should be scattered and not bunched, if possible, in order to prevent spread to the entire flock.

Remove foreign objects—Catch and examine sheep whose eyes are watery, to see if foreign objects in the eye are responsible for the trouble.

Treat with modern drugs—Pink eye, because of new germ killing drugs, is no longer as serious a malady as it used to be. However, it is important to treat the eye immediately with a modern germ killer like the sulfa products. Commercial companies manufacture proper drug emulsions that soothe membranes and are highly effective against the causative agents. In severe cases of pink eye, injection of the product may be necessary.

11. Controlling Blackleg

Two very similar diseases, blackleg and malignant edema, are highly fatal and rapidly progressive in sheep. Losses generally occur after docking or castration or from other factors which may cause wounds. If blackleg is not a problem in your district, prevention by vaccination will not be necessary.

Cause—Two closely related microorganisms are responsible for the disease. Outbreaks are most prevalent during spring and fall pasture seasons.

Symptoms—The disease is easily recognized by high fever, loss of appetite, and listlessness. The most important characteristic is gas-filled, tumor-like swellings that crackle when pressed, and are found around neck and flanks.

Prevention—Sanitation and good management help; however, vaccination is the only reliable preventative measure. Most reliable companies have excellent vaccines available that give at least one year immunity when used according to directions. If an outbreak occurs, it may be necessary to revaccinate.

Destroy dead animals—Burn completely or bury deeply in quick-lime all carcasses so as to avoid contamination with germs that may live for years unless destroyed.

12. Controlling Bloody Scours (Coccidiosis)

While the germs responsible for this disease may be present in healthy, adult animals, it most often affects young lambs. The disease can spread widely through a flock or just affect a few animals, depending on the species of parasite concerned. It is economically important, since animals lose weight rapidly and die, or should they recover, frequently fail to make satisfactory gains.

Cause—California Extension Circular 130 says the disease is caused by a small organism known as a protozoan often found in the feces of normal animals. Favorable moisture and temperature conditions cause it to reach the infective stage when ingested.

Symptoms—Persistent, bloody scours are, of course, the most obvious symptom. Loss of appetite and great depression follow after which the sheep become anemic, thin, and weak. Death may occur a week after symptoms are first noted.

Prevention—Because of the fact the germs may always be present in feces of healthy animals, proper sanitation is the key to prevention. Healthy, well-fed lambs in warm, dry quarters, or well-scattered lambs on

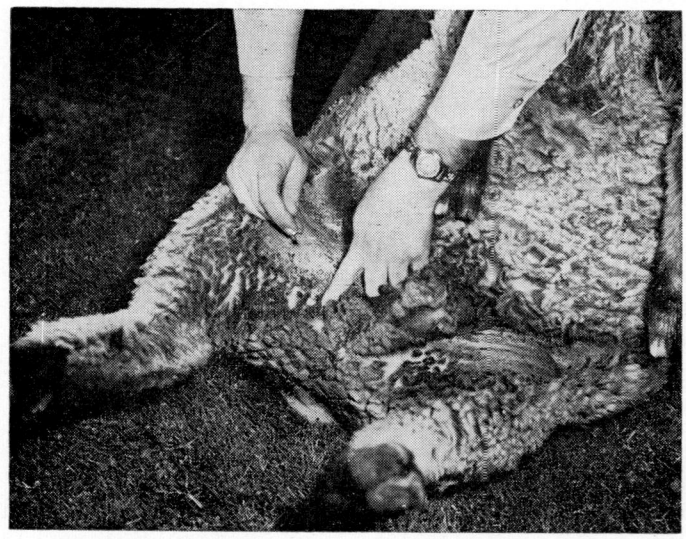

Fig. 8.15—Many minor diseases affect sheep. This sheep is being vaccinated for soremouth (contagious ecthyma) by the operator placing one drop of the vaccine in a scarified area of the skin. (Courtesy, Cutler Laboratories, Berkeley, California)

a good range, greatly cut down the chances of infection. Many sick lambs will recover by themselves if simply moved to warm, dry quarters. It is a desirable practice to separate, if possible, all affected lambs and their mothers from other healthy sheep. Hay should be fed from racks so that droppings will not contaminate the feed.

Most experiments indicate that one to four treatments at 24 hour intervals with sulfa compounds are effective in controlling the disease.

13. Controlling Stiff Lambs

One of the challenging features of sheep husbandry is that there are a number of diseases which are caused by relatively unknown factors. Good management and proper sanitary precautions generally reward the careful operator. One of the factors with which the sheepman must contend is a condition known as stiff lambs. This may accompany several diseases and may indicate a more obscure disorder. Many intestinal, respiratory, urinary, and nervous disturbances may interfere with and cause stiffness of the limbs.

Cause—The true cause is unknown as yet. However, arthritis, posterior paralysis, and muscular dystrophy (white muscle disease) are all characterized by a stiff, awkward gait. Posterior paralysis may be found just after docking which causes an infection in the spinal cord paralyzing the hind quarters.

Prevention—No specific preventive is known; however, proper management, particularly with regard to sanitation and adequate disinfection of navels, docking, and castration wound, will reduce the chances of lambs becoming infected. Farm flocks are more susceptible than range sheep, probably due to the fact sheep are crowded and confined to infected quarters. As a general rule, it rarely pays to attempt to feed out sick lambs unless a ewe would be left without a lamb.

14. Preventing Losses From Poisonous Plants

Poisonous plants are the cause of heavy losses annually among all kinds of domestic animals on pastures, and particularly on range land. Stockmen have recognized the cause of such trouble, yet relatively few are familiar with the poisonous plants found in their area. Therefore, it is important that sheepmen learn to know such plants and put into action approved practices that will eliminate or reduce their losses. **University of California Experiment Station Bulletin 593** suggests a number of precautionary measures to reduce losses.

Learn to recognize primary species—While a large number of plants known to be poisonous exist in any given area, generally there are only a limited number that cause death and serious loss to sheepmen. The larkspur, St. Johnwort, death camas, lupines, and water hemlock are of first importance. Others, such as milkweed, azaleas, laurels, and even tobacco can be factors. Oftentimes one or two species that are not particularly troublesome in general will give a local stockman real concern because of the prevalence of this species on his farm or range. It is desirable, then, to learn to recognize and identify the poisonous species in your district and especially on your farm.

Eradicate toxic species—Where practical, this is an excellent method of preventing loss. Grubbing out and cutting are frequent methods used. Chemical sprays and occasionally soil sterilants are used in well-defined patches.

Fence off danger zones—Frequently growth is so thick, particularly in bogs and swamps, that eradication is too costly. Fencing off such areas is the only practical solution. In time, favorable plants may crowd out the toxic species if no grazing occurs.

Drain wet areas—Bogs may favor the growth of undesirable species and prevent growth of good grasses. Therefore, if possible, spread excessive moisture by draining to other dryer areas as this procedure will help remove conditions favorable to toxic plants.

Herd animals slowly—When moving herds from one range to another, they should be moved slowly. Sheep tend to select wholesome plants if moved quietly. Otherwise, they will nip anything within reach if rushed through an area. Hungry animals should never be driven through an area where poisonous plants abound.

Graze with proper livestock—Various kinds of livestock select different species of plants to eat. In addition, they vary in their susceptibility to its poison. It is good procedure to graze a range with the type of livestock unaffected by the plants growing in that locality. For example, cattle are rarely poisoned by death camas, whereas sheep losses are heavy. Sheep on the other hand never seem to be poisoned by larkspur, and cattle frequently have serious losses.

Provide ample salt and minerals—Any nutritional deficiency tends to give livestock a perverted appetite which may cause them to eat poisonous plants they would normally leave alone. Sack salt is best because it can be distributed so that many animals can eat it at once. In addition, it is more convenient to mix with other minerals like iodine or bonemeal should such elements be lacking in the natural feed or water. It is a good idea to check with local authorities and other stockmen to see what deficiencies exist in your locale.

Avoid too early seasonal grazing—Many plants are

more toxic during one stage of growth than another. Due to this fact, and also because many poisonous plants make a fairly good growth before beneficial plants, early seasonal grazing is a particularly dangerous time. Once the desirable grasses and other plants have reached the proper stage, sheep can get a better "bite" and thus do not ingest poisonous plants accidentally along with normal grazing.

Examine hay carefully—This is especially important when purchasing hay grown outside your own district. Knowing what plants are poisonous and examining hay to see if any such material is contained therein is highly recommended. Checking to see where the hay is produced (especially native hay) may assist in knowing what to look for as some areas are notorious for certain toxic plants. In case of doubt, samples of hay should be sent to the nearest agricultural college, state department of agriculture, or the local extension office for an evaluation of its poisonous weed content. If one suspects danger, it is a wise idea to feed only a few animals (not most valuable ones) first for a week or so to see if there is any effect on them. Occasionally, the same practice of grazing a few head can be tried out on a pasture or range before turning in the entire flock.

Halogeton poisoning a threat—Halogeton poisoning continues to plague livestockmen who graze sheep on the desert regions of the Intermountain West.

Lynn F. James, Agricultural Research Service, explains that halogeton is an annual plant that grows in the colder arid and semi-arid areas. It grows in places where the soil has been disturbed or where the existing plant cover is inadequate or weakened by overgrazing. Therefore, the density of halogeton varies from an occasional plant on the better ranges to increasing amounts on the ranges in poorer condition. Near pure stands can be found along roadways, abandoned fields,

CONTROLLING PARASITES AND DISEASES

old bed grounds, water ponds, around ant hills and in other areas where the soil has been disturbed.

Toxicity results from oxalate—The toxicity results from a chemical compound called oxalate. The amount varies from 10 to 30 per cent of the dry weight of the whole plant.

Halogeton is grazed to some extent all the time by sheep using desert ranges. Under certain conditions, they are enticed to graze excessive amounts which then cause death.

Sheep should be managed carefully on halogeton ranges, with the following practices being observed:

1. Sheep should be introduced slowly onto ranges infested with halogeton.
2. Sheep should not be allowed to become hungry prior to moving into areas having heavy stands of halogeton. This can be prevented by providing adequate forage and water.
3. Supplementary feed should be provided if native forage is not available in adequate amounts. Such should include a mineral supplement as dicalcium phosphate.
4. A grazing plan should be made out so as to avoid grazing sheep on heavy stands of halogeton, such as old bed grounds, around watering ponds and other denuded areas. If sheep are to be grazed into these areas, precautions should be taken.
5. Sheep should be provided with a continuous smorgasbord of forage species.

Chemical Poisoning

Poisoning is fairly rare in domestic animals and often hard to prove except by costly methods. Invariably, it stems from accidental causes, such as leaving empty sacks or bags around that have contained poison that, consequently, is eaten by sheep. In such cases, however, death is swift and sure, and may result in tremendous financial loss to one particular owner who was negligent in allowing such conditions to exist.

Compounds of arsenic, lead, nitrates, as well as copper sulfate, strychnine, etc., are some of the chemicals involved. These are important because they are also commonly used in ordinary farming operations for one reason or another.

Treatment of poisoned animals rarely is practical or timely enough so that precautionary measures are the only effective practices to consider. Old paper bags, sacks, and similar hazards should be burned and disposed of so sheep can never get to them.

Guard against letting sheep graze a field that has recently been fertilized or sprayed chemically, especially from the standpoint that discarded fertilizer containers may be left out for them to get to and thus be accidentally poisoned.

It is a desirable practice to call a veterinarian if the cause of poisoning is not immediately discovered.

15. Controlling Scrapie

Scrapie is an infectious disease that attacks the nervous system of sheep.

The name "scrapie" describes a main symptom of the disease—an infected animal scrapes off patches of wool as it rubs against objects to relieve intense itching.

Usually, scrapie kills animals that show scrapie symptoms. No treatment is effective; no vaccine is available to protect animals against the disease.

United States Department of Agriculture Leaflet No. 457 gives a very complete explanation of the disease and recommends eradication procedure.

All breeds of sheep are reported susceptible to scrapie, but some breeds are more resistant to infection than others. In the United States the disease has been diagnosed only in Suffolk and Cheviot breeds.

Cause—It is believed to be caused by a virus. Scrapie usually spreads from infected flocks to healthy sheep through sale or loan of breeding animals. Little is known about natural spread of scrapie; it may be spread by carrier animals or by other means.

Symptoms—Scrapie is hard to detect. As symptoms develop, an infected sheep is abnormally nervous. It develops slight muscular tremors and its wool becomes dry and lusterless.

Usually, itching starts around the rump and spreads to the back, flanks, and shoulders. An infected sheep shows signs of scrapie when it rubs against trees, corrals, fences, and other stationary objects. It may bite the skin on its belly and legs or pull out wool with its teeth.

The sheep gradually loses weight, although it eats normally. It grows weak and loses muscular control. It walks with an incoordinated gait—swaying, staggering and stumbling frequently.

Generally, infected animals die three weeks to six months after symptoms appear.

Prevention—The only known way to stop the spread of scrapie is to prevent exposure of sheep to the disease. Isolate sheep that rub their wool or show other signs of scrapie. Notify your state or federal veterinarian at once. (Your own veterinarian may contact government officials if he has examined an animal with scrapie symptoms.)

Do not destroy suspect sheep—If scrapie is diagnosed, state-federal quarantines may be placed on flocks that might contain animals with scrapie. Livestock officials locate these flocks by tracing each animal in the infected flock back to its source farm and by checking the movement of all exposed sheep that have left the infected flock. Quarantines are maintained until exposed sheep are disposed of.

Also, all exposed sheep are slaughtered. If your sheep have been exposed but do not show symptoms of scrapie, you may get a permit to take them—under supervision—to an authorized slaughterhouse.

Exposed sheep include:

Sheep in the flock with the infected animal.

Sheep in the flock that was the source of scrapie infection.

Sheep that were in contact with the infected animal before it developed symptoms.

Offspring of all sheep in these three groups.

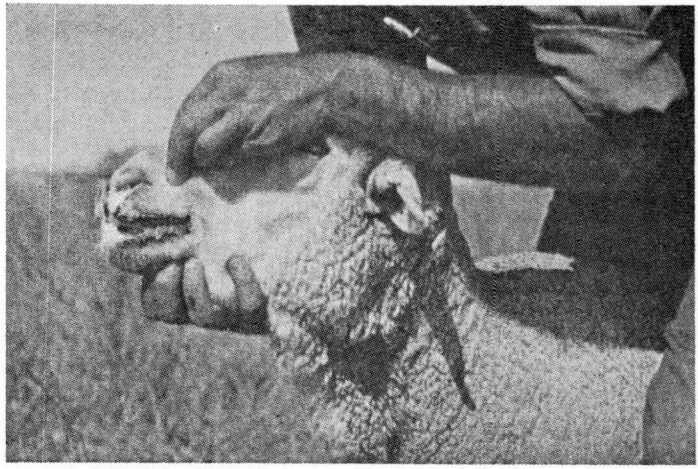

Fig. 8.16—Bluetongue can be accurately diagnosed through laboratory tests; however, typical symptoms are the nasal discharge and inflamed inner surfaces of lips. Not all signs of bluetongue appear in a single sheep, or even in a single outbreak. Usually, the first symptoms are a twitching of the lips, drooling, a watery nasal discharge, and a high fever.

16. Controlling Bluetongue

Bluetongue first was found in the United States in 1948. During the next 12 years, it occurred in Arizona, California, Colorado, Idaho, Kansas, Missouri, Nebraska, Nevada, New Mexico, Oklahoma, Oregon, Texas, and Utah.

Bluetongue, a seasonal disease of sheep, is caused by a virus. The disease is native to South Africa, but has spread to the United States and countries in the Mediterranean Sea area. The name describes the characteristic discoloring of the tongue that occurs in some severely infected sheep. Animals with mild forms of bluetongue seldom develop this symptom.

An infected animal may become weak and emaciated. It may have a swollen muzzle, inflamed, raw areas in the mouth and nose, and high fever.

In the United States death losses range from one to thirty per cent of infected sheep. Yearlings are often affected; however, sheep of all ages and all breeds are susceptible.

An excellent summary is given in **United States Department of Agriculture Leaflet No. 461.**

Cause—Bluetongue, a seasonal disease of sheep, is caused by a virus. It is not contagious; it does not spread by contact between infected and susceptible sheep.

The virus must get into a sheep's bloodstream before bluetongue can occur. A susceptible sheep can be expected to develop the disease within seven to ten days after the virus is introduced.

Scientists believe that the disease may be transmitted by bloodsucking or biting insects—probably certain species of Culicoides, commonly called gnats, sandflies, or no-see-ums.

Symptoms—In northern areas bluetongue season

extends from midsummer until killing frosts. In mild climates bluetongue may occur at any time of year. Not all signs of bluetongue appear in a single sheep or even in a single outbreak.

Usually, the first symptoms are a twitching of the lips, drooling, a watery nasal discharge, and a high fever.

As the disease progresses, a swelling of the lips, ears, and throat frequently occurs. The skin over the face, ears, throat, and flanks—especially of white-faced sheep—becomes red. Sometimes the skin over the entire body reddens. Linings of the mouth and nose first appear intensely red, but gradually change to a bluish-red. Sloughing off of small areas leaves raw, bleeding surfaces in the lining of the mouth, on the tongue, margin of the lips, at corners of the mouth, and in the nose. The tongue may swell, become bluish-red, and protrude from the mouth.

Some sheep with bluetongue become lame. Often the coronary band—sensitive skin margin from which the horn or the hoof grows—becomes inflamed. A distinct red stripe one-eighth to one-fourth inch wide appears along the coronary band; then multiple narrow red strips appear in the bulb area of the heel. The red color gradually changes to a deep bluish-red and usually disappears within two months.

Treatment—If sheep show symptoms of bluetongue, a local or state veterinarian should be called as other serious diseases may be confused with it. No satisfactory medical treatment has been found for animals with bluetongue; however, **good care** will reduce the severity of the disease. Most deaths occur within 10 days after symptoms appear. As a rule, sheep with bluetongue recover naturally within 14 days, but sheep that appear to be severely affected recover more slowly.

Infected sheep should not be roughly handled or

driven. On hot days keep them cool and in shade, if possible, because heat and sun aggravate the condition.

Prevention—Vaccination of healthy sheep with bluetongue vaccine will protect them against the disease. Where bluetongue exists in an area, the veterinarian may recommend annual vaccination of the entire flock.

When possible, ewes should be vaccinated at least one month before breeding. Bluetongue vaccine should not be given pregnant ewes, particularly during the first 60 days of pregnancy. Lambs suckling immune ewes should not be vaccinated until they are three or four months old.

Controlling gnats and mosquitoes may lessen outbreaks of the disease. It is a desirable practice to apply all sanitary measures such as drying up wet areas around water troughs, filling low areas, or spreading manure in an attempt to eliminate insect breeding areas.

17. Controlling Milk Fever

Milk fever in sheep does not seem to be as prevalent as it is in high producing dairy cattle. Nevertheless, it does occur and sheepmen should be aware of the nature of the disease.

Cause—It most commonly occurs soon after parturition or lambing. Most authorities agree that the immediate cause is an acute drop in the blood calcium concentrate, although the balance or ratio of other minerals in the blood also seems to be associated with the disease. Naturally, when milk forms in the udder and especially when milked out, the calcium and other elements in the blood are withdrawn so rapidly that they cannot be replaced fast enough and milk fever occurs.

Symptoms—The term milk fever is a misnomer as

the temperature of the animal may, in reality, fall. Loss of appetite and general depression occur, but the most typical symptom is for the animal to eventually go down or collapse with the head pulled back against the flank.

The symptoms of lambing paralysis are similar to milk fever. Dr. Glen Spurlock, Animal Husbandry Department, University of California, Davis, says it is important for sheepmen to understand the difference as diagnosis can most quickly be determined by the results of treatment.

Treatment—Generally, a solution of calcium gluconate is injected intramuscularly. A veterinarian should perform the task unless the producer is qualified. Treatment is almost always successful, and the animal will recover in a short time if the cause is **milk fever**, but this will **not** be the case if the ewe has lambing paralysis. Later injections may be necessary. Feeding a calcium-rich ration should prevent recurrence of the condition.

Good milking ewes would probably be most apt to get milk fever. Therefore, it would be desirable to attempt to save them, in addition to determining whether the cause is lambing paralysis or milk fever.

CHAPTER IX

BUTCHERING LAMB AND MUTTON ON THE FARM

Lamb is one of the great meat delicacies, yet only a relatively small amount is eaten by rural families. Most of this is probably due to food habits, as farm families who try it invariably find its characteristic flavor a real treat when this convenient fresh meat is added to their regular diet.

The small size of the lamb particularly adapts it for farm slaughter. Lambs generally dress out around 50% so that an 80 pound lamb would yield approximately a 40 pound carcass. The entire carcass is, therefore, small enough so the ordinary household refrigerator can protect the meat from spoilage until it is eaten, even in the summer.

Some persons prefer mutton to lamb because it has a more mature flavor. However, whichever one prefers, the approved practices necessary to properly kill and dress the carcasses are the same.

Activities Which Involve Approved Practices

1. Killing the lamb.
2. Removing the pelt.

3. Removing the viscera.
4. Caring for the pelt.
5. Cutting up the carcass.

1. Killing the Lamb

The kind of meat that eventually will be served on the table depends to a large degree on the kind of lamb selected for slaughter. Thrifty, well-finished lambs make the most desirable carcasses. While 80 pound lambs are commonly used, many persons prefer lambs weighing 90 to 100 pounds or more.

Withhold feed—Animals should be kept off feed for 10 to 12 hours prior to killing. However, they should have free access to water. Be certain that the fleece keeps dry, as it taints the meat easier when damp.

Keep animals quiet—Rough handling, especially grabbing sheep by the wool, causes bruises and makes an unattractive carcass. Bruised meat not only is unsightly, but has poor keeping quality. Handle with one hand under the throat and the other under the dock. If no one is around to hold the lamb, the two front legs may be tied to one hind leg and the animal laid on its side.

Assemble equipment—Don't leave butchering preparations until the last minute. A sharp knife, preferably a skinning knife, a steel, and a saw are the principal tools necessary. In addition, six feet of clean, quarter-inch rope is needed with which to hang the lamb to a stoutly anchored two x four beam or tree limb about seven feet from the ground. A bench or clean floor upon which to lay the lamb, a tub for offal, a pan for heart, liver, and tongue, plenty of water, and a few clothes complete the list.

Stick with sharp knife—A sharp, long, thin-bladed sticking knife is best. Grasp the lamb by the lower jaw

BUTCHERING LAMB AND MUTTON ON THE FARM

Fig. 9.1—A modern slaughter house is a highly automated industrialized operation. However, sheep are killed in a humane fashion.

with one hand, and force a knee against the top of the head, thus stretching out the neck. Sticking is generally accomplished by pushing the knife through the throat just back of the jaw bone with the cutting edge towards the backbone. Both arteries and veins are thereby severed and death quickly follows. Letting the animal struggle promotes thorough bleeding. Some persons prefer to stun the animal by breaking its neck after sticking. This can be done by pushing sharply down on the head while simultaneously pulling back on the chin.

Wash the dead animal—Not many persons butchering sheep follow this practice as it is considered unnecessary with clean sheep, and too time consuming. However if the animals are very dirty or there is danger of getting wool grease on the meat, it may be desirable to wash the body with clean, cool water **before** skinning.

Shear first—Animals killed at home for family use

are handled and skinned easier if they are sheared first.

2. Removing the Pelt

Cut off head—After bleeding, place the lamb squarely on its back and rip the skin down the center of throat to the point of jaw. Skin out and unjoint head at the first joint. Remove the tongue and if planning to use, remove the brain also. Wash in cool water.

Skin brisket first—Generally, the pelt adheres to the brisket more tightly than other spots, so it is a desirable practice to skin this out first. It may be necessary to use the knife occasionally and even to slit the skin in a few spots in order to get the skin loose. Pull back

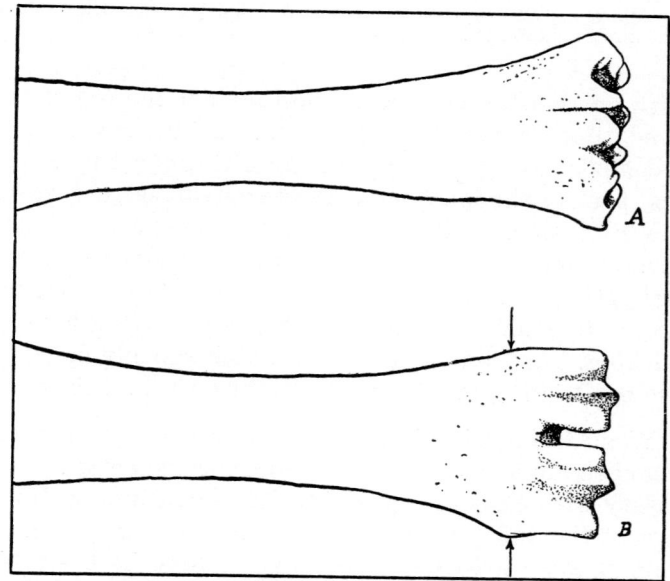

Fig. 9.2—The two types of joints of the foreleg; A, the false or "lamb joint" and B, the true "mutton joint" of a sheep. Arrow indicates the location of the lamb joint or suture.

on the pelt and skin out both sides of the neck as far as possible.

Skin for shank—The fore shanks are ripped down the back and skinned down just below the knees.

Score and break leg joints—The joint below the knee is scored and broken. This may be tricky to locate at first, but a little practice makes the breakjoint easy to find. A common error is to score too high up on the leg. Round joints indicate mutton; square break joints, lamb.

Fist pelts backwards—This operation refers to using a tightly clenched fist to work loose the pelt from the body. The fist is worked all around the body, loosening the pelt without cutting it. In this fashion, the "fell" is unbroken. Wash hands frequently during this process. When the pelt is all loosened over the belly, it should be ripped down the center.

Saw breast bone—Using the saw, split the breast bone while the animal is still on its back. After this, split the throat down the center and trim out the windpipe and gullet as far back as possible.

Skin hind legs—Stretch the hind legs forward and rip the pelt covering them from a point back of the hock to the area just back of the udder or cod. Unjoint the back legs low enough so as to preserve the tendons for hanging the carcass. Skin out the hind legs.

Fist pelt forward—Fist the belly forward from the rear in the same fashion as the brisket was fisted backwards. Rip the skin so the pelt lies open.

Hang carcass—At this point, the entire carcass may be suspended so as to keep it clean and make the remainder of the skinning process easy. Tie the hind legs together loosely by placing a cord through a slit made between tendon and leg bone and securing so the carcass may be suspended.

Some prefer to use a gambrel hook to spread the hind legs a little. This makes it much easier to remove the bung and gut, and to split down the bottom of the pelvis.

Fist remainder of pelt—Work upwards over the legs and loins and downwards over the shoulders. Be sure hands are kept clean so the carcass will not have a "woolly" taste. Furthermore, it is advisable to pelt and clean the carcass as quickly as possible after killing if a choice carcass is to be the result, as gases form quickly inside the body once the animal is dead.

Loosen tail and pull pelt from back—The rest of the pelt is removed by using the knife to work around the tail and then pulling the entire pelt down from the back and off the carcass.

3. Removing the Viscera

This operation is easier and cleaner if a box or can is available for the unwanted part of the viscera, and an ample supply of clean water is close at hand.

Rip belly wall—Make a short incision line through the upper belly wall. With the knife pointing out and towards the operator, insert the clenched fist holding the knife into the opening and slit the entire belly down to the throat opening. The fist prevents the knife from cutting any of the intestines.

Remove organs—While the kidneys and kidney fat are left in the carcass, the liver, heart, and lungs are removed along with the entire viscera. Place the liver and heart in a clean pan. The diaphragm must be cut through in order to remove heart and lungs. Cut through diaphragm where red and white flesh meet.

Wash carcass—Blood in the vessels along the back should be massaged out by working from center toward flanks. Wipe out any soiled places and wash the entire carcass.

Chill carcass—After carcass is clean, it should chill for about 24 hours. Meat should not be frozen immediately as the flavor will be impaired. Before chilling, the tail should be pinned down and the fore flanks pinned back. At home or on the farm, carcasses may be hung outside at night providing nights are cool and the carcass is protected from rain. In the day it can be wrapped in blankets; however, cutting up the carcass is much easier if the chill is sufficient to solidify the meat. If the meat is to be stored frozen in a home freezer, many people freeze the meat first before cutting it up with a power meat saw. It is too difficult a task to cut frozen meat by hand so that firm chilling is best only when power saws are unavailable.

4. Caring for the Pelt

Lamb and sheep pelts are of sufficient value so care should be taken in their handling and disposal.

Sell pelts green—When only one or two pelts are involved, as well might be the case on the average farm killing lamb for home consumption, the easiest and least bothersome way is to sell the hide immediately to a local hide dealer. This should be done the first or second day, as they deteriorate rapidly. Prices are not as high this way, but the difference in price is not great enough to merit the extra time and labor needed to cure and store.

Air dry hides—If hides are to be stored or cannot be sold immediately, they should be air dried. This can be accomplished by hanging over a wall or fence flesh side up and in a well ventilated spot. Spread out the pelt as much as possible so it will dry flat and quickly. Another way is to rub the flesh side with fine salt and leave it to cure. Put plenty of salt near the edges. Pelts will cure in 15 to 30 days. Pelts should always be stored fleece side down.

5. Cutting Up the Carcass

If the lamb has been properly killed, chilled, and a good meat saw and sharp knife are available, it is a relatively simple task to cut up the carcass. A good, sturdy, wooden table will be a real asset. There are many different ways a lamb carcass can be cut up, depending on how it is to be used. However, the following method is recommended for general home use.

Omit splitting—In general, a lamb carcass is cut up like any other meat animal with the exception of not splitting the carcass down the back into two "sides." The only exception to this, on the farm, is to split the carcass immediately as an aid to cooling.

Cut according to guide lines on chart—The following chart shows where the basic cuts should be made in cutting up the carcass.

It is a desirable practice to cut up the carcass in the order as shown below:

- a. **Remove shoulder.** Saw through entire carcass between fifth and sixth ribs. Cut off the neck at the top of the shoulder and remove the shanks. Then separate right and left shoulder by sawing through backbone.
- b. **Saw off breast piece.** Start with a knife and finish with a saw. It is easiest to do if the carcass is on its side and legs toward you.
- c. **Cut off rack.** Do this by cutting between last two ribs. After splitting it may be cut into rib chops.
- d. **Remove sirloin.** As this is a small roast, it can be

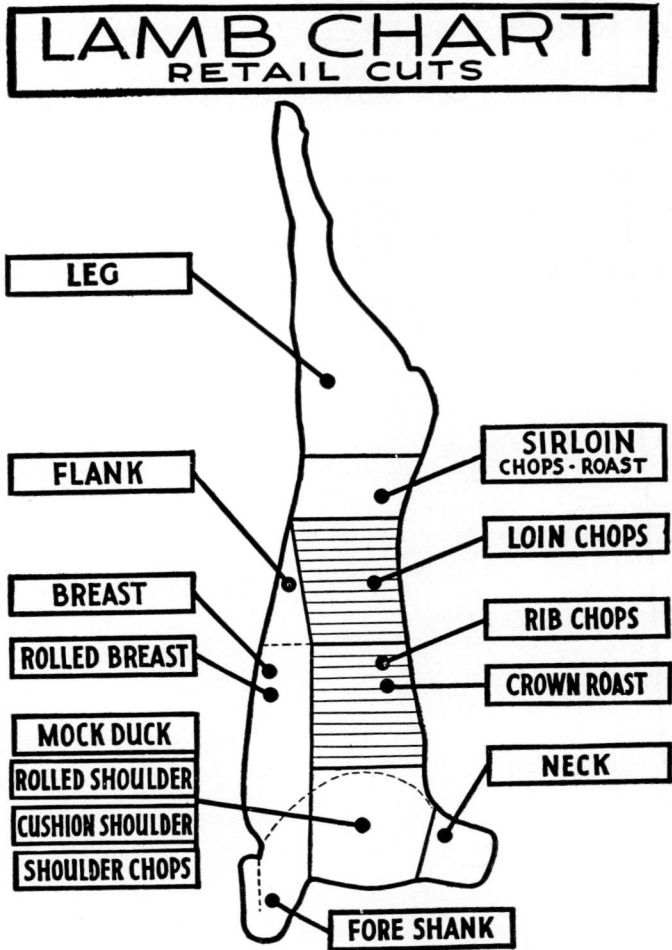

Fig. 9.3—Carcasses are cut into wholesale cuts as indicated by heavy, black lines.

cut off at any preferred thickness. Use both knife and saw.
e. **Separate loin from leg.** Make the cut at the small of the back or at the pin bone. Cut loin chops as desired.
f. **Split lamb legs.** Separate the legs into two equal parts with the meat saw.

The entire carcass is now ready for cooking, canning, curing, or it can be wrapped and frozen. It may be a good idea to trim off tag ends so the cuts look neat and appetizing.

CHAPTER X

SELECTING AND USING LAMB AND MUTTON

The selection of lamb is generally not as painstaking as that of beef as there seems to be less of a grading variation and, therefore, less chance for the inexperienced to make mistakes. On the other hand, it is an advantage to know what one is looking for in good lamb and which cuts are most suitable for every occasion.

Activities Which Involve Approved Practices

1. Selecting lamb and mutton.
2. Cutting and trimming cuts for cooking.
3. Freezing lamb.
4. Canning lamb.
5. Curing lamb.
6. Smoking lamb.

1. Selecting Lamb and Mutton

Knowing what to look for in meat is the key step in trying the most desirable cuts.

The United States official grades, in their respective order for the different kinds of meat, are:

Lamb	Mutton
Prime	Choice
Choice	Good
Good	Utility
Utility	Cull
Cull	

According to **Farmers' Bulletin 1807**, the desirability of lamb as a meat depends upon four main factors. They are the breeding or breed of the animal, how and what it has been fed, the age of slaughter, and the method of handling. Therefore, it is advisable to know the history of what you wish to eat. While impossible with most store cuts, it is obviously one of the advantages of home-killed lamb.

Use mutton breeds—Lambs of mutton breeds are usually more suitable for eating than those of wool breeds. Such animals fatten readily, dress a higher percentage, and have meatier roasts and chops.

Learn difference between lamb and mutton—The term "mutton" as commonly used commercially, applies to meat of older sheep, that is, ewes over 12 months of age, and wethers over 18 months of age. While mutton is excellent meat, it is generally less tender than lamb. The common way of distinguishing between mutton and lamb is by the break joint on the knee. When scored and broken, the break joint of lamb is square and rough, while mutton is round, smooth, and a short distance down the leg.

Determine the amount of finish—A well finished lamb yields a carcass that is fairly well covered with fat over the legs and shoulders as well as the back. However, a high degree of finish is not essential for home use. Lambs that have made rapid gains and are moderately fattened produce carcasses of tender, high-quality meat.

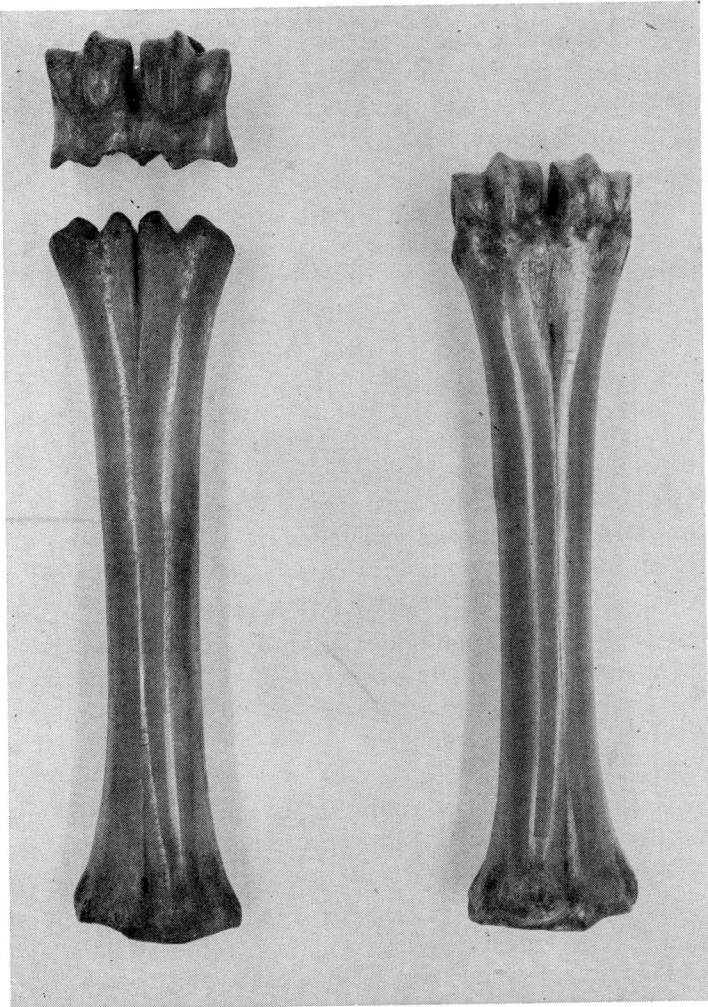

Fig. 10.1—Characteristics of lamb (left) and mutton (right) carcasses. This illustrates the metacarpal bones of sheep showing a typical break joint, depending upon the age of the animal.

2. Cutting and Trimming Cuts for Cooking

The major portion of lamb is eaten as fresh meat, either as roasts or chops. More efficient and desirable meals will be served if each part of the lamb is utilized in the most appropriate way.

Learn to identify cuts—Each part of the carcass is, of course, best utilized when cut and served in its proper manner. Home-killed meat has a more appetizing appeal when cut up in an approved fashion. Therefore, the first step is to be able to identify the following cuts:

Shoulder—This, of course, is the meat from the upper part of the foreleg. It is cut either standard cut or long cut. Neck and shank are removed. A long cut shoulder is used in making a boneless shoulder roll by removing neckbone and ribs from shoulder and unjointing shoulder blade and arm bones. When rolled and tied this makes an attractive long cut, boneless shoulder which may be used fresh or cured.

Neck—While easy to identify, this cut is often neglected. It may be sliced three-fourths to one inch thick and used for stewing, braising, or cooking in a casserole, or the bone may be removed and the slices stuffed with sausage.

Breast—This cut is best served by removing rib ends, breast bone, and then rolling it. Start with shank, roll tightly, and tie with stout cord. The breast cut may be roasted, cured, or canned.

Shank—Shanks are used for stewing, meat broth, or small shank roasts. Remove shank from shoulder cut whenever square cut lamb shoulders are desired.

Rack—The rack may be served in a variety of ways. A pleasant appearing crown roast or rib chops can be made from the rack. For rib chops, split the rack lengthwise and make chops by slicing through the ribs.

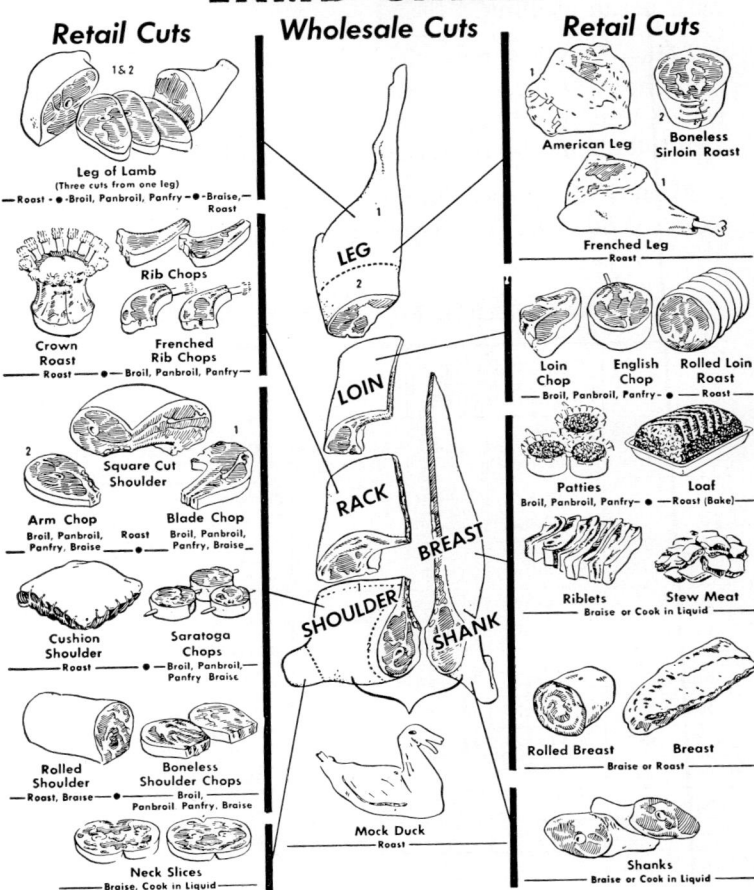

Fig. 10.2—Courtesy of National Live Stock and Meat Board.

A crown roast is made by trimming out backbone of a strip of meat several inches wide from the ends of the ribs. Use six to eight ribs. Bend rack and mold into shape by tying the two sets of ribs together. When cooking, wrap rib ends with salt pork and fill crown with bread stuffing.

Loin—This is a most desirable cut and may be sliced into chops of any thickness. It may also be boned and rolled, tied tightly and roasted, or else cured or canned. The fell should be removed before cooking.

Flank—Many possible ways of serving the flank exist. When cut into small pieces, it is excellent for

TIME-TABLE FOR COOKING LAMB

CUT	ROASTED AT 300° F. OVEN TEMPERATURE		BROILED		BRAISED	COOKED IN LIQUID
	Meat Thermometer Reading	Time	Meat Thermometer Reading	Time	Total Time	Time
	Degrees F.	Minutes per lb.	Degrees F.	Minutes	Hours	Hours
Leg	175 to 180	30 to 35				
Shoulder Whole	175 to 180	30 to 35				
Rolled	175 to 180	40 to 45				
Cushion	175 to 180	30 to 35				
Breast Stuffed	175 to 180	30 to 35			1½ to 2	
Rolled	175 to 180	30 to 35			1½ to 2	
Lamb Loaf	175 to 180	30 to 35				
Chops (1 in.)			170	12		
Chops (1½ in.)			170	18		
Chops (2 in.)			170	22		
Lamb Patties (1 in.)				15 to 18		
Neck Slices					1	
Shanks					1½	
Stew						1½ to 2

stewing. If ground and added to other pieces, it makes choice lamb patties.

Sirloin—After the sirloin is removed from the rack, it can be trimmed by removing backbone and hip bones. This piece can then be rolled, tied into shape, and the result is an excellent roast. The sirloin may be cured also. Many people prefer simply to slice the sirloin into chops along with the loin and use these cuts as frying pieces.

Leg—The choicest part of the lamb is considered by many people to be the leg. It is meaty, has excellent flavor, and is a favorite for many who frequently make it the main course at a Sunday dinner. After trimming, it is generally roasted, although when cured its appearance resembles a ham. The trimming may be used to make lamb patties; adding one-third pork to the lamb improves the patties as pork tends to prevent crumbling when cooking. The leg is served either French style or American style.

American style has most of the bone removed and will fit into a smaller pan or oven. Slit down the side of the leg bone and remove the bone at the knuckle joint.

French style is made by scoring around the leg about two inches above the joint. Next, cut over the break joint just above the hock and break it over a table edge. Then, twist the joint until it comes free from the leg bone.

Consume early—While storing in the cooler at 34° for nine or ten days will ripen meat and improve the flavor, it should not be kept too long. Whenever lamb is to be stored longer than two or three weeks, the chilled storage of a refrigerator or pantry will rarely be satisfactory. It must either be frozen, canned, or cured.

3. Freezing Lamb

This method has become increasingly popular with everyone since the arrival on the market of practical home freezers. According to **Farmers' Bulletin 1807**, frozen lamb can be kept satisfactorily for six months or longer; however, frozen meat tends to dehydrate somewhat so that it is a desirable practice to buy or kill an amount that will be consumed by this time before adding more lamb meat to the locker.

Trim cuts as they will be served—Meat is hard to trim just after thawing. Therefore, it is wise to trim all cuts, prior to freezing, exactly as the housewife will use them.

Use proper wrapping material—Only moisture-proof paper should be used. Various manufacturers sell proper wrapping paper for freezing meat and no substitute should be used, as poor paper, tape, etc., will work loose or else moisture will seep through. One of the most desirable ways of wrapping is to purchase new-type plastic bags and simply put cuts of meat inside. The advantage of plastic bags is that meat can be seen and identified through the bag. In addition, they can be cleaned and used over again.

Wrap tightly—The so-called "drug store" wrap is the best way in which to package meat. Use care in

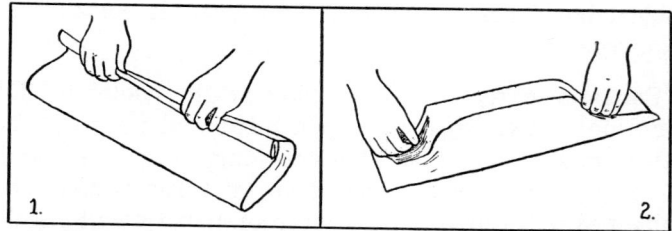

Fig. 10.3—The drug store wrap as sketched is a good way to wrap meat for home freezing. Another excellent way is by the use of clear, plastic bags, as they may be used repeatedly.

fitting paper tightly against meat, especially along the exposed surfaces of the lean. It is a desirable practice to try and eliminate as much air as possible from the inside of the package while wrapping. Some persons use plastic bags for wrapping and exhaust all air by means of a vacuum cleaner before sealing. Sealing out the air keeps the meat more moist and, hence, better textured for later use.

Package according to family size—The number of chops per package or size of roasts, etc., should be in keeping with the size of the family or the contemplated use. By this procedure, only one package need be taken from the refrigerator at a time in order to prepare a meal. Waste is also prevented, as too much meat is often thawed unless packages are the right size.

Label promptly—As soon as the meat is wrapped, it should be promptly labeled as to the kind of meat and the cut. For example, Lamb—Loin chops. Some prefer to place the number and even size of cut on the label. Colored labels denoting kind of meat are excellent.

Freeze immediately—This prevents moisture seeping through the package. In addition, it prevents loss of moisture in the meat by freezing before the meat sweats.

Keep a running inventory—Whenever meat is added or taken from the locker, a record should be made. In this manner, the housewife knows just what is available for use and avoids needless piling and unpiling of packages. Searching for certain cuts keeps the freezer open too long and causes unnecessary frosting on the inside of the box.

Freeze quickly—A temperature of 0° or —10° F. is probably better than higher ones for freezing lamb, although temperatures of 10° F. are used successfully for storing frozen lamb.

Cook promptly after thawing—Thawed meat is likely to spoil more quickly than fresh meat; therefore, it should be cooked promptly before bacteria have a chance to develop and cause spoilage. Chops and roasts will cook quicker and more uniformly if thawed first. However, stew meat may be put into the kettle without previous thawing. For best results, thaw meat in the kitchen refrigerator rather than at room temperatures.

4. Canning Lamb

According to **Farmers' Bulletin 1762**, a satisfactory method of preserving lamb is by canning. This way is not used as much as it previously was before home freezers came into general use. However, canning results in a tasty product and has the advantages of easy transportation of meat and almost indefinite storage.

Can quality meat—Use the same meat that you would for fresh cuts. It is important to cut away any blood clots or bruised areas, as poor flavor often results from these areas.

Make sanitation a "must"—Tables, knives, and utensils must be absolutely clean. Meat must be wiped with a damp cloth, if soiled.

Prepare meat for canning—Meat may be prepared for canning by one of four methods. It may be canned raw, seared, boiled, or cured before canning, depending upon the taste desired and the amount of preparation one cares to do when the time comes to eat.

Destroy all organisms—All meat must be canned in a steam pressure cooker to insure perfect safety. Other methods of canning are not satisfactory as meat is more difficult to can than fruit. Either tins or jars are satisfactory.

SELECTING AND USING LAMB AND MUTTON

Store properly—Cans or jars must be cleaned and stored in a cool, dry place. Mark containers to show contents and date of canning.

Examine carefully before eating—Cans which bulge or leak indicate spoilage. When opened, there should be no outburst of gas or liquid. Watch for a peculiar odor or moldy appearance. **DO NOT** taste or eat food that appears to be spoiled, and canned meat generally should not be tasted until after heating.

Farmers' Bulletin 1762, or Morton Salt Company's booklet, **Home Meat Curing Made Easy**, will give additional information on canning.

5. Curing Lamb

Lamb is easily and quickly cured; however, after smoking, the cuts dry rapidly and tend to become somewhat strong in flavor. Many persons prefer this somewhat "gamey" flavor, and each year smoke a few pieces for special occasions. One advantage of curing lamb is that it is a relatively safe procedure and can be used wherever refrigeration is lacking.

Legs cure best—While any cut may be cured, the leg of lamb is the best piece and will keep its agreeable flavor longer. It should be cured a day or two after slaughter or even before when it is fresher.

Cure dry or brine method—Both methods are satisfactory. The backbone should be split and the spinal cord removed, as it spoils easily.

Dry method—For each 100 pounds of meat use five pounds salt, four pounds sugar, and four ounces saltpeter. Mix thoroughly and rub into each piece of meat and on the bottom of a clean container. A crock or hardwood barrel should be used. Use two-thirds the mixture the first time and the remainder three to five days later when the meat is overhauled. Meat should remain in the cure about one to one and one-half days per pound after which the surplus salt should be brushed off and the meat wrapped in parchment paper. A light washing in tepid water is advocated providing the meat is allowed to dry thoroughly. Store in a dry, well-ventilated place. It is a desirable practice to mix some of the cure with water and pump an ounce or so of the mix into the joints, using a pumping pickle plum.

Brine method—This is a very satisfactory method of curing any kind of meat. A standard formula is eight pounds salt, two pounds sugar, and two ounces saltpeter dissolved in six gallons of clean water. This is enough to cover 100 pounds of meat. Pack the meat into a crock or wooden barrel and pour in the brine. Use a board and stone to hold meat down as all meat must be covered by the brine. Overhaul meat on third to fifth day. Add more brine if necessary. Thin cuts will be cured in two weeks whereas five or six pound legs will take 30 to 40 days. Best results are obtained when the brine is pumped into the joints with a pump prior to pickling. Many commercial companies (for example Morton's Tender-Quick) sell ready-mix cures that are very satisfactory. It is a desirable practice to own a pump to pump in the brine as this generally eliminates all chances of spoilage around the bones and knuckles.

6. Smoking Lamb

Scrub thoroughly—The cured lamb should be scrubbed with hot water and hung up to drain and dry before smoking.

Use hardwood smoke—This will vary in accordance with the country in which one lives. However, oak, hickory, and most other hardwoods have top smoking qualities. Hang meat far enough away from the fire so that the smoke and not the fire reaches the meat.

Smoke until colored—Generally two days at 100° to 120° F. will properly smoke lamb. It may be left in longer if desired, but it also tends to dry out if left in too long. It is best if the meat is eaten within four to six weeks.

CHAPTER XI

MARKETING MUTTON, LAMB, AND WOOL

The marketing of farm products is an area to which farmers traditionally have given little attention. Many producers will spend months of hard labor, risk their capital, and yet give little thought or study to the best method of disposing of their hard-earned produce. It is true that in many instances there is little that can be done by farmers to influence market prices; yet, with the large volume involved, in most cases a fraction of a cent per pound is a significant gain or loss for the farmer. In addition, other factors are very important in influencing market prices within a given range. Quality or grade of product, seasonal difference in prices, method of shipping, selection of type of market, even marketing according to weekly or daily price differentials are some of the factors that producers have a degree of control over or, at least, can use to best advantage in disposing of their products.

Sheep raising is in many respects the most speculative of all livestock enterprises. Wool prices may vary tremendously from year to year, even to the extent of going from a very high price to a "no market" situation in a matter of months. Cooperatives for marketing wool have a long record of failure, yet there are some valid

Fig. 11.1—Miss Margie Sharp, Miss Wool of America 1971-72. As most production problems can be solved, marketing is the bottleneck to higher profits. The Miss Wool contest is one activity used to stimulate the use of wool.

reasons for their existence. In view of all these factors, it is imperative that sheepmen use approved practices in marketing their crop if they are to derive the most profit from the business.

Activities Which Involve Approved Practices

1. Marketing lambs.
2. Marketing mutton and other sheep.
3. Preparing wool for market.
4. Marketing wool.
5. Shipping livestock.
6. Preparation for show or sale.
7. Conducting a sale or auction.
8. Photographing sheep.

1. Marketing Lambs

In this chapter we shall not consider any of the productive practices which must be employed in order to produce top quality, well-grown lambs. However, it goes without saying that marketing is an easier task with quality lambs, and top prices are received only for desirable animals. Therefore, it is important that all approved practices are utilized in order to produce a high quality product, and we shall consider marketing as the activity which takes place after the lamb is ready to sell.

Get market price—No one expects to sell any great number of animals for more than market price, yet oftentimes producers will let their livestock go below its true value. It makes no difference how you sell your lambs as long as their full value is received. You may sell to local buyers, speculators, through a commission firm, to a central market or direct to a packer. Experience will help in this activity, but it pays to shop around and see that your lambs are bringing their true value.

Keep posted on prices—It will pay you to visit a terminal market and observe how prices and grades are determined. The state wool growers' association, magazines, newspapers, and radio are all excellent sources of price information. Prices must be current, so one must be careful to get latest prices as they can change rapidly. It is a desirable practice to telephone buyers or commission firms in advance so that lambs can be brought to market at the best time. Commission firms appreciate advance notice and frequently a delay of a day or two may make a significant gain in price.

It is a wise idea to cultivate the acquaintance of several buyers or dealers on more than one market so that when sale time comes you will be in a position to place your lambs on the market most favorable for you.

Deliver only to known buyers—Lambs should not be delivered to unknown buyers except on receipt of cash on hand on delivery. Commission firms should be those which are approved on regularly established markets. When selling lambs for future delivery, be sure you know the buyer has a good reputation and require a contract with an appropriate deposit.

Market at peak condition—Milk lambs, in particular, are a perishable product. Therefore, it is important that sheepmen watch carefully the lambs' development as they approach saleable age and market them when they reach maximum value. A few days is a safe tolerance if one is waiting for certain changes in market conditions, favorable shipping conditions, etc.; however, once lambs reach their peak bloom, they should be disposed of as quickly as possible. There may be good and bad sales, but producers should consider the average in judging their success. Do not hurry into a decision, but do sell before it is too late.

Sort lambs—Sorting will result in a desirable quality of lamb going to market, which in the long run will build a reputation for the producer and put his lambs in demand. Lambs should be fat, of good quality and weigh between 80 and 90 pounds for most buyers, although heavier lambs may sometimes mean greater profit providing feed is no problem and lambs are in demand. When lambs' weights reach 110 pounds, buyers will generally discriminate against them by discounting the price. Wether and ewe lambs may be sold together, but ram lambs, off weights, thin lambs, etc., should be sorted and sold separately as they tend to bring over-all prices down rather than raise poor lambs close to the price of the good ones. Frequently, if top lambs are sorted out and sold, the weaker ones will pick up and eventually develop into heavier weights. It is a desirable practice to attempt to have a fair-sized lot of lambs for market at the same time, as buyers never give as much attention to one or two head as they do a large lot. Range producers and those with many animals seldom have this problem; however, farmers with small farm flocks should try to get the number offered for sale to at least 20 head or more.

2. Marketing Mutton and Other Sheep

Lambs, of course, constitute the largest and most valuable product offered for sale by the sheep enterprise. On the other hand, a large assortment of the various grades, weights, and ages must periodically be disposed of which, in the long run, will make up a sizeable part of the sheepman's income. Old ewes, cripples, rams, large wethers, feeders, culls, etc., are a portion of what sheepmen may offer for sale from time to time. Of this group, feeders are perhaps the greatest in number and should be given the most attention; however, most sheepmen consider them in a similar category to fat lambs and market them accordingly, the only difference being that the ewe lambs may be kept for replacements while lighter weights and those in poorer condition are sold as feeders.

With the advent of irrigated pastures, the demand for feeders has been stepped up, so it is important that this group be marketed to best advantage.

Most of the approved practices listed for marketing lambs apply to all sheep, but in addition, the following practices should be observed.

Cull early—A dead ewe has little sale value. Cripples, poor feeders, off-type animals seldom, if ever, make satisfactory or economical gains. For this reason, it is important to dispose of the animals as quickly as possible in order to eliminate further feed costs or hamper the orderly management of the rest of the flock. Occasionally, small flock owners will be confronted with the problem of culling old ewes. Even though animals are obviously culls from the standpoint of age and condition, farmers may keep such animals, pointing out the fact that they bring little when offered for sale and yet might produce a healthy lamb. When such is the case, it is important to decide early if such a calculated risk is to be undertaken and then either sell as quickly as possible or do everything possible to maintain the animal's health.

Consider local auctions—While it may pay to ship feeders to distant points or central markets, oftentimes the best policy for cull animals is to dispose of them as close to home as possible. Decrepit sheep are more apt to die en route or become crippled so that most sheepmen like to truck cull animals to local auctions and take whatever value they can get. Freight costs are less and very often such small numbers are involved that it does not warrant a long haul.

Feed out through custom feeders—Another possibility is for sheepmen to send their feeders off their own premises to custom feedlots. Here the sheepman has the opportunity of having his animals fed out by someone else and of making an increased profit because his lambs will grade higher when offered for final sale.

3. Preparing Wool for Market

Wool must be properly prepared and carefully handled to bring full value when marketed.

Attractively prepared wool can usually be sold at a premium over wool in poor condition.

Marketing Bulletin 10 of the United States Department of Agriculture says that:

"Proper shearing is important—**growers should supervise it carefully.** Other points vital to good preparation are keeping the wool clean and dry, tying fleeces attractively, and packaging them properly.

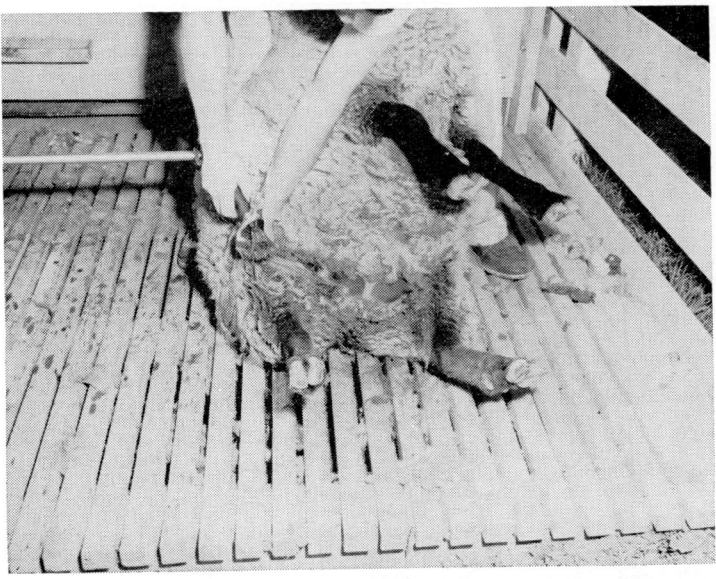

Fig. 11.2—A desirable floor on which to shear sheep. The slats allow dirt and chaff to fall through and give good footing on which to work. Smooth, hard lumber is best.

"The manner in which wool is prepared for market has a great influence on the quality and value of the finished cloth—hence on the marketability of the grower's clip."

Keep sheep dry—Protect sheep from wet weather during shearing. Wool that is packed when damp will become discolored during storage. Wet wool requires special handling and drying by handler or dealer. Do not pack wet wool.

Texas Agricultural Extension Service MP-244 summarized the approved wool production practices as follows:

Summary
APPROVED WOOL PRODUCTION PRACTICES*

Practice	Recommendation	Good	Fair	Needs Improvement
1. Practice tagging or crutching.	Tagging or crutching should be done before lambing and to prevent fleece worms.			
2. Provide a clean place to shear.	Clean the pens and wet them down.			
3. Keep mohair out of wool.	Clean mohair out of shearing pens.			
4. Shear on shearing boards or floor.	Hire a crew that shears on boards and has its own shearing boards.			
5. Use trip boards.	Keep trash and dirt off shearing floor.			
6. Separate black sheep and sheep with black faces.	These sheep should be shorn last and the wool packed separately.			

(Continued)

Summary
APPROVED WOOL PRODUCTION PRACTICES (Continued)

Practice	Recommendation	Good	Fair	Needs Improvement
7. Use new lubricating oil.	Request shearers to use new lubricating oil or supply new lubricating oil for shearers.			
8. Give the shearing crew definite instructions.	Use proper shearing, rolling and packing methods.			
9. Keep the shearing floor clean.	Keep tags picked up, the shearing floor swept clean, and pack tags separately.			
10. Roll fleeces properly.	Roll with flesh side out.			
11. Tie fleeces properly.	Use paper twine only and not excessive amounts.			
12. Pack bags uniformly.	Pack bags either flat or round; the average weight should be 180 to 200 pounds.			
13. Keep bags clean.	Lower bags onto tarp or clean floor after packing.			
14. Practice grading.	Practice only when warehouse is cooperating.			
15. Haul in a clean truck.	Clean truck or trailer; cover with a tarp.			
16. Store in a safe place.	Keep clean and dry, preferably where insured.			
17. Keep records.	Keep records of average shearing weights, price, shearing price and bag weights.			

* From Texas Agricultural Extension Service MP-244

Fig. 11.3—Fair managers who, for years, have been trying to dress up livestock departments may tear a page from Playboy, or other national publications carrying ads for Pendleton Woolen Mills.

In the ads are sheep—in PLAID COLORS.

Western Fairs Association has long advocated such gimmicks as colored sawdust and floriculture displays in judging rings, special pens to tell the story of individual breeds, better stall cards, cleaner aisles, and other accouterments of barn showmanship—thus the interest in Pendleton's dressed up lambs.

The formula: (1) Sheep are washed with detergent and water, after it has been determined that they are not allergic to the detergent. (2) Water-based vegetable dyes are prepared. (3) A stencil is held tightly against the wool and the dye is air-brushed on. Zig-zags are made by shifting the stencil. (4) If a light base is required, the sheep should first be dyed all over and allowed to dry before applying the plaid. (5) For laborious designs, such as zig-zags, it might be well to first tranquilize the sheep. (Courtesy, **California Livestock News**)

4. Marketing Wool

While the actual selling of the wool crop happens only once a year, its preparation and making ready for sale is really a year-round job. **Circular 171,** University of California, states that the preparation of wool for market should include the entire growth period of the clip. Adequate care and management of sheep in order to grow the best possible fleece are just as essential as putting up the clip in an attractive package.

Use specially prepared branding fluids—Western sheep are generally branded once a year for identification purposes. Professor Wilson, University of California wool specialist, says house paint or other hard-drying oil bases are insoluble in scouring and, therefore, responsible for thousands of dollars worth of damage every year in the textile industry. Sheep-branding liquids must stay on sheep for a year, yet wash out in the normal scouring process. It is a desirable practice, then, that only marking fluids manufactured specifically for such a purpose be used and then as sparingly as possible.

Handle wool as though it were perishable—Because wool is not as perishable a product as some farm commodities, it is often carelessly handled after removing from the sheep. **Illinois Circular 657** recommends the following steps in preparing wool for market.

1. Keep the fleeces free of burs and chaff.
2. Handle the wool carefully at shearing time so the fleece is practically in one piece and is kept clean.
3. Remove the tags from the fleeces and sell them separately.
4. Roll the fleeces with the skin or inside of the fleece on the surface.
5. Tie the fleeces with about seven feet of special

WOOL FIBER GRADES, CLASSIFICATIONS, VALUES & BREEDS By G. C. HUGHES, MONTANA WOOL LABORATORY

GRADE	FINE			½ BLOOD		⅜ BLOOD	¼ BLOOD		LOW ¼ BLOOD	COMMON	BRAID	
American English	80's	70's	64's	60's	58's	56's	50's	48's	46's	44's	40's	36's
.0000 inch		.00082		.00094		.00105		.00124	.00136	.00143	.00151	
Microns	20.8		21.9	23.54	24.8	26.9	30.4	33.0	34.8	36.6	38.3	39.3
DIAMETER Comparative Cross-sections	○		○	○		○	○		○	○	○	○
CLASSIFICATION	Combing 2" and over			Combing 2¼" and over		Combing 2½" and over	Combing 2¾" and over	Combing 2¾" and less	Combing 3 inches and over	Combing	Combing	
	French Combing Between			French combing Between		Combing 2½" and less						
	Clothing 1½" and less			Clothing 1½" and less								
VALUE Combing	$1.27			$1.23		1.13	1.05	0.99	0.95	0.90	0.90	
French Combing	$1.22			1.17		1.08						
Cleaning basis clothing		1.12		1.10								
SHRINKAGE—Comparative relationship	60%	62	64	55	57	51	47	49	42	37	33	
VALUE—Grease basis Boston	50.8	46.4	40.3	55.4	50.4	55.4	55.7	50.5	55.1	56.7	60.3	
BREEDS (*two grades including next coarser)	Rambouillet Merino	Targhee* Romeldale	Panama*	Aust. Merino	Corriedale* Columbia*	Down Breeds	Cheviot Dorset		Romney Marsh	Border Leicester	Lincoln Cotswold	Eng. Leicester

Actual length Combing French Combing (Between Combing & Clothing) Clothing Coarser grades (56's and Coarser) are usually combing length.

326 APPROVED PRACTICES IN SHEEP PRODUCTION

EXPLANATION OF TABLE

1. Diameter measurements are averages used in the classroom for fiber measurement. The measurements are mostly USDA findings.

2. The cross-sections show comparative diameters.

3. The bars in classification represent fibers of the actual length listed. French combing length is midway between combing and clothing.

4. "Strictly staple," not included in the table, refers to lengths of ½ to ¾ inch and more than "combing" lengths. The clean price for "fine staple" is $1.29 and "½ blood staple" is $1.25 per pound clean basis. In Territory wools our "combing" usually sells with the "staple" lengths.

5. The clean values listed are the present quotations used by C. C. C. The U. S. Tariff Commission has revealed by a long time study that clean wool prices show the following comparative relationships: 1. The circled 10 indicates an average difference in value between combing and clothing lengths in the same grade of ten cents per pound scoured basis. 2. The circled seven indicates an average difference in value between adjacent grades of the same classification to be seven cents per pound scoured basis.

6. The comparative shrinkages listed are used for grade comparisons and do not represent breed or area shrinkages. Usually there is 2 to 3% more shrinkage for each shorter classification within a grade from the same band of sheep. As a general rule, the average difference in shrinkage between adjacent grades of the same classification is 4 to 5%. This applies to fleeces from the same band. The finer grade has a higher shrinkage.

7. The longest length in a grade has the highest value and lowest shrinkage compared to the shortest length in the same grade having the lowest value and highest shrinkage.

8. The shrinkage and grease price are no indication of clean wool production per sheep or fleece value.

9. The breeds are listed within the grade of highest percentage produced.

paper wool-tying twine. (Fibers from other twines become entangled in the wool and cannot be removed except by hand during the process of manufacture.)
6. Pack the wool in standard wool sacks which hold about 250 to 300 pounds. Use only new wool bags. Firm sacks, but do not overload.
7. Know what kind of wool the sheep produce and sell it on the basis of its market class and grade.

Sell early—This advice, of course, is not iron-clad and is most applicable to small flock owners. Sheepmen, as a rule, are not set up to hold wool nor do they have the contacts to dispose of their wool through speculation. For this reason, local buyers representing large wool firms are generally about as advantageous to sell to as any. Holding wool may occasionally result in a higher price but often simply results in storage costs or loss through insect damage, or necessitates borrowing money in order to carry on other activities. It is a desirable practice to sell immediately whenever wool is at a fair price or higher. Waiting for a higher price usually is disastrous even though occasionally wool may climb higher before the price breaks. Sheepmen operating on a large scale with a recognized clip may be in a position to bargain, but, unfortunately, the small producer is at the mercy of wool trade and other economic factors beyond his scope. It is doubtful if he would gain by trying to play the market. If wool is very low, as it sometimes is, then a producer may be justified in waiting. However, generally speaking, it is advisable to sell for cash. Some small producers have their wool manufactured into blankets and other products and sell these retail as a means of getting a fair price for their wool.

Livestock producers could do much toward increasing marketing efficiency through use of simple yet sound purchase or sale contracts. Many contracts now in use

LAMB PURCHASE CONTRACT

Contract No. ....

This contract, made this .... day of .... A.D. 19........ between (name) .... of (address) .... party of the first part, the seller and (name) .... of (address) .... party of the second part, the buyer.

Witnesseth: For and in consideration of $..................... representing approximately 15 percent of the contract value as advance and part payment on this contract receipt whereof is hereby acknowledged; the seller hereby sells and agrees to deliver to the buyer, and the buyer agrees to buy from the seller the following described lambs on the following terms and the seller does hereby guarantee the title thereto:

.... head of unshorn lambs, ...% black face and ....% white face bearing .... brand, from about .... ewes, less approximately .... head of ewe lamb replacements, at a price of $.................... per cwt. to be delivered f.o.b. shipping point to .... or to other designated points mutually agreed upon between the .... day of .... 19........ and the .... day of .... 19........ at the option of the buyer; seller to hold lambs at his own expense until delivery date specified. (Seller may consent to hold lambs beyond specified delivery date at buyer's expense or on other mutually agreeable terms.)

Lambs shall be weighed with dry fleece and one of the three shrink options:

(Cancel provisions that do not apply)

1. After 12-hour stand without feed or water, or
2. After .... mile drive without feed or water, or
3. Shrink ....% or ....lbs.

Lambs described above are bought subject to State and Federal inspection and seller guarantees that interstate shipment of same can be made from delivery point. Buyer shall have authority to reject sick, crippled, undocked, or diseased lambs, and all lambs under .... lbs. except that buyer cannot reject more than 2 percent or .... lambs on basis of weight.

Balance of the purchase money to be paid to seller by the buyer at the time and place of delivery as stated above, through acceptable Bank exchange.

If the buyer fails to accept the lambs on the terms herein described, the advance payment made at the time of the execution of this contract will be retained by the seller as liquidated damages for the breach of this contract.

Provided further that in the event parties to this contract cannot agree they shall submit their differences to a board of arbitration consisting of three persons (one to be selected by each party to the contract and the two thus selected shall select a third) whose decision shall be final.

Other conditions (specify) ....

....

.... Buyer

By .... Seller

The undersigned, holder of a mortgage or other lien on the livestock above described in consideration of One Dollar, receipt whereof is hereby acknowledged, and other valuable consideration, hereby consents to the sale of said lambs to .... free of such mortgage or other lien upon the terms and conditions above specified.

....
Witness   Mortgagee

Fig. 11.4—A sample purchase contract. (Courtesy, **Extension Bulletin 211,** Utah State Agricultural College Extension Service.)

fail to meet all of the essentials of a good contract. Actually, a number of "so-called contracts" are not contracts at all but mere "options to buy." Inequities and tremendous losses are suffered each year by both producers and buyers because of faulty contracts or failure to use written contracts.

Pool wool—This seldom has any effect on price, but may result in getting wool to market quicker and more efficiently. It also is a way for very small producers to get enough wool together to justify the attention of a dealer.

Investigate all selling channels—Wool may be sold on commission, through wool pools, by cooperatives, local buyers, or wool merchants. A good agency to handle and sell your wool would be one that could satisfactorily answer the following questions:

1. Is the service rendered efficiently and economically?
2. Does it properly differentiate quality of different lots?
3. Is each lot dealt with impersonally?
4. Can market price be manipulated?
5. Will it establish prices which induce producers to produce what the consumers want when and where they want it?

To date, no one best way has been found and it is up to each grower to decide what agency is best for him. Don't rush your product to market so hurriedly that adequate thought hasn't been given to the best way of selling your wool. However, it is desirable to make a decision reasonably early before buyers have left the field. An old-time suggestion or adage is to select a method of selling and then stick with that method over a period of years.

5. Shipping Livestock

After months spent in careful feeding and management of livestock, all can be lost in a very short time while livestock are going to market. Not only must sheep reach the terminal market alive, but if best prices are to be received they must be in good health, peak bloom, and have a minimum of shrinkage due to shipping. It is important that the following approved practices be observed in order to prevent loss in transit.

Limit feed—Rations should not be altered materially. However, heavy concentrate feeding should be limited or even withheld. Show stock, of course, might get a limited amount if they are used to it, but range stock are best fed only good quality legume hay or native grass. Plenty of water should be available, but salt should be omitted for this period as it causes them to drink too much water. Oats are an excellent feed for lambs in transit.

Truck from range to railhead—This practice is of no concern to producers located close to their market. Under these conditions lambs are best trucked directly from farm to final market place. However, it is a desirable practice to truck lambs rather than drive them as the shrink can be reduced and ewes can be left on the range for a longer period. Individual animals, such as valuable breeding stock, are sometimes crated for shipment.

Rest before loading—After driving sheep preparatory to shipping them, it is wise to let them rest for one or two hours to perhaps a day before loading into cars. Even if they have been trucked for some distance, it is well to rest them for a period rather than immediately load for another long trip. This practice is more important with lambs than with mature sheep.

Fig. 11.5—Learn to handle sheep properly. Here is one way to hold and lead a sheep so it easily can be turned in any direction. The arm is placed under the chin of the sheep so it cannot get its head down and the right hand is under the dock so the dock can be squeezed if necessary in order to control the animal.

Load carefully—The first phase of shipping is loading, and it is particularly important that it be done correctly as this is the first experience animals will have that is quite different from ordinary production activities. For this reason, they are apt to become nervous or excited, and "bolt" with the possibility of injuring themselves or even the attendant. Every ranch should have a well-constructed loading chute to load livestock. If large numbers of sheep are to be handled, the chute should be especially designed to handle sheep, although an all-purpose chute will work. (See Chapter VII.) The important thing to remember is that loading equipment be solid because animals shy away from rickety floors. Urge sheep gently but firmly to load. Never grab the wool. Oftentimes picking up one lamb and putting it inside the truck will entice the others to enter, as they are extremely gregarious and follow blindly whatever the leader or other sheep in the band will do.

Sand floors—Truck beds and concrete ramps should have dry sand sprinkled over them to prevent slipping. In warm weather it may be wet down in order to prevent excessive heat build-up. In cool weather it is generally unnecessary to do anything else for sheep although some operators will cover the sand lightly with straw. It is especially recommended that the bed of the truck be kept as clean as possible to avoid staining of the wool. For this reason only small amounts of sand, water, etc., should be put on the truck bed in order that it does not become muddy. Some successful operators are using redwood for floors and loading chute ramps, as this wood is soft, thereby preventing animals from slipping, and is sufficiently rot resistant to stand up under long usage.

Load proper number per car—Overcrowding is a dangerous practice and may result in many crippled animals or a high death rate in transit. The preceding table is a guide in loading for rail shipment.

Usually, it is better to load fewer head in the upper deck than in the lower.

The same precautions are applicable in loading trucks. A good general rule is not to overcrowd sheep, but a firm load, so sheep can brace against each other while traveling, is better than a loose load.

Any truck larger than a pickup should have one or more partitions to prevent the sheep from piling up

Type Car	Weight of Lambs	Number Per Deck
36' lower deck and single deck	75 pounds	125
40' lower deck and single deck	75 pounds	138
36' lower deck and single deck	100 pounds	105
40' lower deck and single deck	100 pounds	116
36' lower deck and single deck	150 pounds	85
40' lower deck and single deck	150 pounds	94

when sudden stops are made or continual up or down grades are encountered.

Sort for size—While lambs of varying weights can be safely transported together, it is a desirable practice to ship large ewes, weak culls, rams, etc., separately or at least partition off each lot securely within the car or truck.

Insure the cargo—Ordinarily, animals moved locally are not worth the time and expense nor is the risk great enough to warrant insuring the animals against loss in transit unless, of course, very valuable breeding stock is involved. On the other hand, commercial trucking companies as well as railroads insure all cargos, including livestock, as a matter of course. Local insurance agencies are the handiest and most practical to work through or should they not insure livestock themselves, can direct ranchers to agents that do. Terminal markets and commission firms are also good sources of information on acceptable insuring agencies for sheep.

Clear with authorities—Delay in shipping will be eliminated if all officials concerned are cleared through first. With local transactions there are seldom any regulations involved. However, on long shipments, especially interstate, some regulatory measures generally are involved. Brand inspection is almost always necessary at terminal markets and clearance must be obtained before leaving your county. These inspectors are located at terminal markets. The extension agent or state Bureau of Animal Identification can locate the nearest source to your farm. Health inspection is often necessary, particularly when breeding stock is involved and animals are going from show to show. County veterinarians occasionally are employed to do the service free although most producers prefer to call their local veterinarian.

Guard against hardship—Bruising, crippling, etc.,

are all risks to shipping. However, the greatest risk is shipping fever. No satisfactory prevention has been worked out but exposure and hardship seem to be the most important predisposing factors. As far as shipping is concerned, adequate rest will do much to keep the resistance of animals at a high level. If lambs are dipped, they should be dried off well before shipping.

Insist on frequent loading and unloading—The federal "28 hour law" says that livestock **cannot** remain on cars longer than 28 hours without being unloaded for feed, water, and rest, except when the shipper signs a release that it may extend to 36 hours. Livestock must remain off the cars for at least five hours. Under certain conditions and where ample room is available, livestock may be fed and watered without unloading. These same general rules should apply when ranchers are transporting their own livestock with their own trucks, as an outbreak of shipping fever can be disastrous. When trucking, drivers should concentrate on steady, easy driving to see that animals do not get down. Sudden stops and quick turns are dangerous. Wind and rain that cause undue exposure should be protected against by the use of straw, canvas, and other protective coverings. A desirable practice is to consult weather reports in advance of shipment so that transportation can take place under most favorable weather conditions.

The use of antibiotics and vaccines to protect animals is still somewhat controversial. Some livestock men swear by these products; others consider them of no value. Many producers take the attitude that such products can do no harm, therefore, why not use them "just in case." Unless a producer has strong feelings on the subject based on his own personal experience, it is advisable to consult the local veterinarian and follow his recommendation.

Avoid overcrowding lambs in car—Experienced men avoid overcrowding lambs in the car. About 140 to 150, 60-pound feeder lambs can usually be loaded in a single-decked car or about 300 head in a 36 foot double-deck car. The exact route over which the lambs are to be shipped should be carefully designated in advance. While trucking has many advantages, lambs stand rail shipment better.

6. Preparation for Show or Sale

Many questions may arise as to why it is necessary that sheep be fitted or groomed in preparation for show or sale. It is true that the vast majority of sheep are marketed with no formal preparation. However, even in these cases, some preparation actually takes place. For example, lambs that arrive at market properly sorted, free from mud, injury, wool pulled out, etc., always make a better showing and command full market price. Therefore, the main idea in preparing animals prior to sale is not to make them appear better or look like something they are not, but to make them present their true value to the buyer. In this manner, the lamb is not penalized because his good qualities are hidden. When animals are fitted for show, the degree of fitting, of course, is much greater than when preparing for sale. Purebred producers can gain in the long run by participation to some extent in shows or fairs for these reasons:

1. Exhibiting is an excellent means of advertising.
2. It provides an opportunity to sell surplus stock.
3. There is an excellent opportunity to study market and breed types as well as compare your sheep with others.
4. Occasionally, premium money may be won.
5. New ideas are gained from others.

MARKETING MUTTON, LAMB, AND WOOL

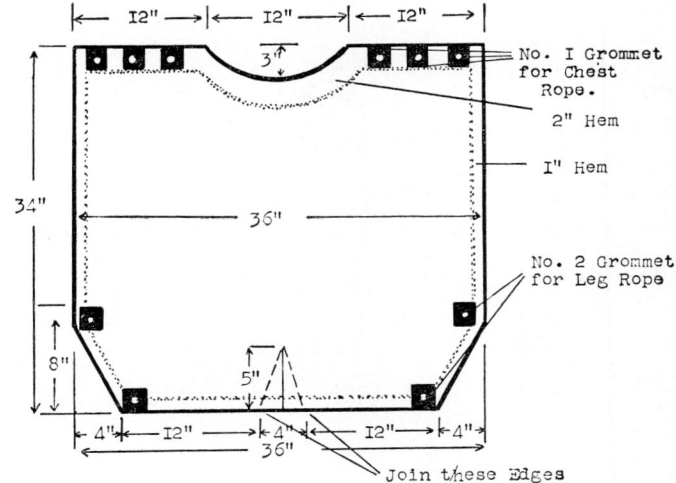

Fig. 11.6—Sheep coat pattern. (Courtesy, Wool Department, University of Wyoming)

The main disadvantage is the extra time, effort, and expense incurred in showing and preparing to show. Therefore the degree of fitting will depend upon what the final purpose is to be—a sale would require a minimum of fitting, whereas a large fair or show, considerably more. Whatever the purpose, the approved practices involved are much the same.

Start fitting early—Good feeding practices, feet properly trimmed, freedom from parasites, and other approved practices necessary for general sheep production are basic to properly fit show animals. Additional practices must also be employed and the timing of these practices is important. Fat lambs going to a fall show should have all the wool sheared off in June when lambs are started on feed. Then in August, shear the wool off the top of the back to make the back flat on top. With spring lambs that are milk fed, do not shear the entire

Fig. 11.7—A group of ewes wearing sheep coats. Sheep coats assist in keeping the animals clean and improve the appearance of the wool.

body, but do block sheep well in advance. A general rule to use to begin fitting is:

1. Shear 8-12 weeks prior to show for young sheep.
2. Shear 4-6 weeks prior to show for aged sheep.

Quarter properly—Sheep must be comfortable. Good, clean pastures are fine. Barns should be roomy and cool in the summer. Good ventilation and clean bedding are essential. If sheep are allowed to become stained or dirty, much additional work is required to get them clean.

Trim the feet—Appearance will be enhanced and crooked feet prevented by proper foot trimming. First, set the sheep on its rump, lean it back against your legs with the head hanging over left thigh. Another method is to lay the sheep flat on its side.

MARKETING MUTTON, LAMB, AND WOOL

Then, using a sharp knife trim the excess hoof off the underside of the foot and shorten the toes. Do not draw blood. Excessively long hoofs may have to be trimmed with a shears first. Show animals have this done about once a month.

Prepare the fleece—The first step is to assemble all tools so the operation will be speeded up as much as possible. A suggested tool list for preparing the fleece is:

1. One or more hand shears.
2. A stiff brush and clean cloth.
3. A halter and bucket.
4. Pair of wool cards.

For a fall show, most showmen like the procedure

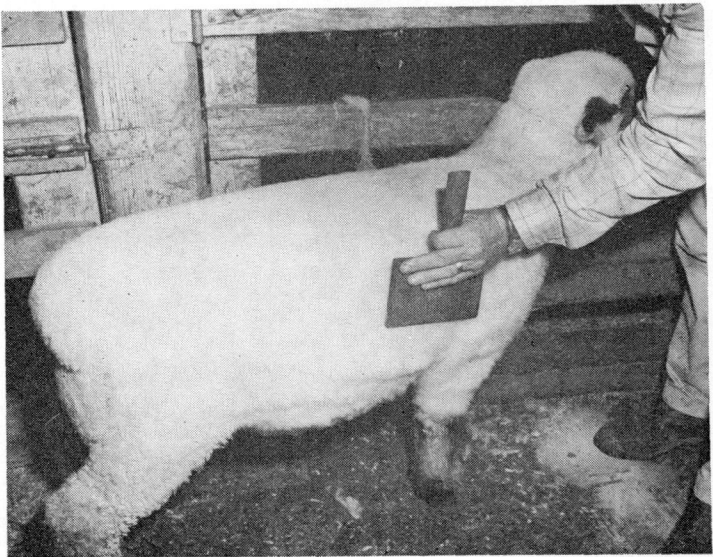

Fig. 11.8—Illustration of the proper way to hold and card the wool on a sheep. Sheep must be clean and fitted; however, in modern shows and exhibitions not as much wool is left on as in previous years.

Fig. 11.9—Proper method of trimming a lamb for show. The shears are held fairly level and flat against the wool while the blades are activated in a fairly rapid motion. Most sheepmen prefer to take a small cut and then go over the animal several times. (Courtesy, Robert Finley, Herdsman, University of California, Davis)

somewhat as follows although many people have individual preference in regard to exact dates:

1. Shear in March and shear close.
2. Dip the last of May.
3. Clip back down level and square dock in June.
4. Clean and card the fleece and trim it smooth all over about August 1.
5. About a week before show time, trim again and put on a light blanket. Some sheepmen blanket much before this time. (Gunny sack will do.)
6. Trim the fleece just before show time.

The use of sheep dip in about one-half pail of water will make carding and trimming easier and give a more uniform appearance to the fleece. (Use just enough sheep dip to give a deep brown color.)

If sheep become unusually dirty, they should be dipped six weeks prior to the show as this will give the grease time to get into the fleece.

Trimming sheep properly is a matter of experience and the beginner should try not to become discouraged too quickly. A desirable practice is to observe and get aid from the county agent, teacher of vocational agriculture, or an experienced sheepman before attempting to trim or block an animal.

Precautions for competitive showing—Listed are seven practices which should be followed when showing at a fair or show, although each individual producer will have his own technique.

1. Check entries to avoid error—see that health

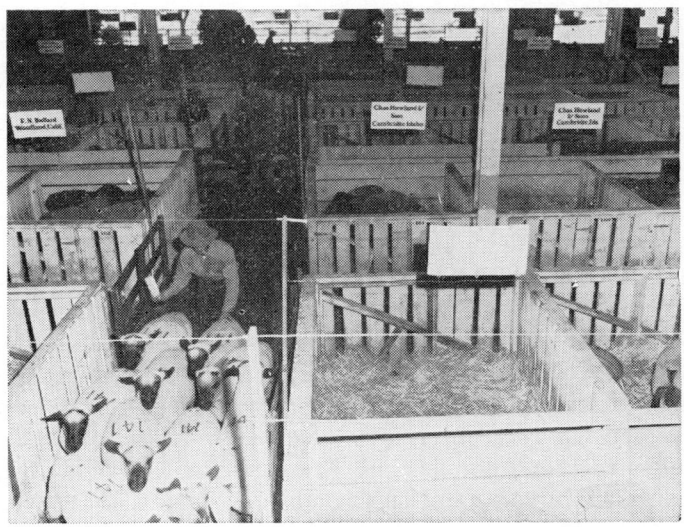

Fig. 11.10—Scene at a well-conducted sale. These rams are being brought up rapidly so they will enter the ring on time and not keep buyers waiting. All sheep must be properly and easily identified.

certificate, registration papers, etc., are in order. (Follow the premium book for each show.)
2. Have animal ready to lead in when called.
3. Be sure **you** are ready—dress for the occasion.
4. Dampen fleece slightly and pat with card to show bloom.
5. Remove dirt from face or muzzle with damp cloth.
6. Fluff hind quarter to make it look fuller.
7. Remove straw from underline and legs.

Rules when in the show ring:

1. Stand sheep squarely on its legs and never let the judge catch it out of position, if possible. Keep your eye on the judge.
2. Push against brisket with knee causing sheep to tense muscles of its back so the back will be firm to touch.
3. Stand on left side of animal. Hold sheep with left hand under jaw leaving the right hand free for other movements and to move the animal when necessary.

7. Conducting a Sale or Auction

A well-conducted sale, especially for purebred breeders, is an excellent method of disposing of good livestock. Quality of animals being offered is the key factor in holding a successful sale. Sheep must be offered that you would be willing to keep in your own flock. Common practice in offering rams is to hold an auction whereby many producers submit and offer for sale a few top animals. This is an excellent method both from the buyers' and the sellers' viewpoint. Whatever method is used, there are a number of approved practices to follow.

Plan well in advance—Any sale should be planned

far enough in advance to properly prepare your animals and advertise the event. This takes months, not weeks, of preparation. Publications, fieldmen, association secretaries, etc., will be able to adjust their schedule and attend if advanced notice is given.

Get the best auctioneer and assistants—The best auctioneer will more than earn his fee as people trust his judgment and respect his reputation. Planning well in advance will insure getting a good auctioneer, as qualified ones are busy. Do not overlook good local auctioneers as their knowledge of local tradition and local demands can be put to sale advantage. Get extra help for the event. Your neighbors are your best bet, as they are sincere in their desire to help and know your layout in advance.

Offer only attractive animals—Animals should show evidence of being handled and properly fitted. Have their feet trimmed and be sure they are in peak bloom. Group animals for easy inspection, especially before the auction, so buyers can make a preliminary inspection. Have your herd books open and available for inspection.

Offer only healthy animals—Sheep should be free from all external parasites. Dip if necessary, well in advance and clean the premises. If internal parasites are suspected, animals should be properly treated. The best breeders will guarantee the health of their livestock and make good any violation.

Advertise well—The amount of advertising will depend to a large extent on quality and number of animals to be sold. Livestock journals, market papers, local and weekly newspapers, as well as radio, are good mediums of advertising. Do not neglect your neighbors, as they may turn out to be your best customers. If a sale catalog is necessary, 400 to 500 copies are adequate. Pictures will help. Be honest in all advertising and insert no misleading concepts. Let the 4-H, FFA, and other junior groups in on your sale—they are good present and excellent future customers.

Insure the comfort of the group—Make the buyers feel "at home" and as comfortable as possible. Provide a pavilion for shade or heat as necessary. Well lighted sale quarters and good ventilation will go a long way in keeping buyers at the sale. Get animals into the ring **on time.** Do not keep buyers waiting. Be certain the ring is arranged so everyone can easily see all animals being offered. Lunch, rest rooms, and parking facilities should be provided.

Complete transfers rapidly—When the sale is over, get the transfers of ownership and other papers out as soon as possible and in a business-like fashion. Follow up on your animals and show an interest in their future progress. Previous buyers are your best source of future sales.

Stand behind your animals—Make a fair adjustment of all non-breeders, etc., whenever necessary. A slight immediate loss will more than repay with future sales because buyers gain a feeling of confidence when they know that breeders guarantee all animals offered for sale.

8. Photographing Sheep

One of the best methods of advertising is by the proper use of good pictures of livestock. Catalogs can be made more effective if they include photographs of animals being offered for sale.

Records are more meaningful if they contain pictures. There are a number of approved practices to be followed in obtaining good photographs. If the animals are valuable enough, it might pay to have a professional photographer, skilled in photographing sheep, take the pictures. However, many additional factors must be considered even then.

Proper fit the animal—Sheep should be prepared for a photograph just as though they were going to enter the show ring. Remove manure stains.

Proper clipping is important. Trim sheep's feet so they will stand squarely.

Good grooming will pay, as it brings out the best points of an animal. Sheep must be in good health and well fed if they are to be photographed.

Fig. 11.11—A well-fitted animal properly posed for photographing.

Select the proper growth stage—Animals should be in good condition and healthy. Select a time of year when the wool is long in order to get the best picture. Ewes look best just before lambing. Young, growing animals often appear better one month than the next.

Avoid distracting background—Select a site where

telephone poles, old machinery, etc., will not be visible. A lawn is a good location. Landscapes, such as a low, rising hill or hedge, make an ideal setting and do not detract from the animal. Feet should be squarely placed and head high.

Use right amount of straw—Many pictures can be taken on bare, clean ground or lawn with no straw whatsoever. Sensible amounts of straw may dress up and improve pictures in some instances. In no case should photos be taken on dirty concrete or in messy corrals. Do not bury animals in straw, as buyer may become suspicious.

Post the animal—Make certain that the animal is alert by waving a hat or in some way attracting its attention. Young animals should have the two feet nearest the camera well apart. A little grain scattered on the ground will often pose or "group" animals correctly for photographing. One good trick is to pick up the forequarters or the entire animal several inches off the ground and drop them. In most instances, the animals will land squarely on their feet and in proper stance for picturing.

Do not take pictures in the wind—Focus directly on the animal. The picture is slightly improved if the background is not sharp. Take the picture a little above the animal; never from a lower or kneeling position. Some animals may appear better on one side than on the other.

Keep records—Age of the animal, stage of lactation, dates, and location are important if the photograph is to be used for most purposes.

Use only good pictures—Patience will be rewarded. Do not retouch pictures. If the picture does not turn out well, discard it. A poor picture is worse than none.

CHAPTER XII

RECORDS FOR THE SHEEP BUSINESS

There are some sheep producers who keep no records whatsoever and yet are able to remain successfully in the sheep business, at least to some degree. Others may keep only a checkbook stub or a record of actual cash and expenses ignoring for the most part any records concerning their livestock. On the other hand, most successful and progressive sheepmen consider it a desirable practice to keep an adequate set of books both on their animals and on the economic aspects of the business. Good records will prove helpful in determining proper income tax reports, in comparing one season's profits to the next, in disposing of surplus animals, in planning future breeding programs, and a host of other valuable information to guide the business ventures of each sheepman. Good records can tell you what is wrong and what is right with your business. It is not necessary that records be long and complicated; quite the contrary. The best records are neat, simple, precise, and contain only those items necessary for proper operation.

Total value of United States cattle and calves on farms and ranches January 1, 1972, was a record high $24.6 billion, 16 percent above a year earlier. The value per head averaged $209, up $24 from a year earlier. The

LIVESTOCK VALUE, JANUARY 1 AVERAGE PER HEAD
IN UNITED STATES, 1970-72[1]

Class	1970	1971	1972
		Dollars	
Cattle and calves	180.00	185.00	209.00
Stock sheep and lambs	24.70	23.50	22.60
Hogs and pigs[2]	36.98	23.37	28.50

1. Source: **USDA Statistical Reporting Service.**
2. Previous year December 1 values.

sheep and lamb inventory was valued at $423.5 million, down 9 percent from a year earlier. The value per head for sheep and lambs was down 90 cents to an average of $22.60.

Cattle values increase, sheep and hogs off—When livestock prices are down, it may be a good time to enter the industry, as price structure tends to be cyclical. However, many other factors, like personal income, eating fashions, or relative cost of production figure must also be considered. In any case, records are needed to provide the basic data for decision making.

Activities Which Involve Approved Practices

1. Determining kinds of records to keep.
2. Keeping financial records.
3. Keeping production records.
4. Analyzing records.
5. Establishing credit.

1. Determining Kinds of Records to Keep

It is important that records be kept simple and in such a form that they can be understood by anyone qualified to interpret livestock records. The more involved records are, the more one can tell about his business providing, of course, they are properly analyzed. The best kept records in the world mean nothing unless they are summarized at the end of the season so comparisons can be made to previous years and other breeders.

Keep track of costs—Regardless of how unimportant an item may be, every producer should keep a record of all expenses. Some producers write checks for all purchases and use the stub as an expense record. While this is better than no records, it has the disadvantage of not itemizing purchases so they can be charged against the correct enterprise. A record of all income must be kept, but this is generally not a critical point as it happens only once or twice a year in the sheep business. Home consumed products and gifts should be listed as income to the enterprise.

Production records—These can be simple, but no progress in a breeding program is possible unless sheepmen know what production animals are making in relation to feed, type of sire, etc.

Transaction records—Bills of sale, transfer of ownership, registration, and similar records are very desirable for the purebred producer to keep. Such information is valuable to himself and also instills confidence in buyers who wish to purchase animals from his flock.

2. Keeping Financial Records

The importance of financial records is best emphasized by the fact that many producers keep financial reports as their only record. Entries showing items of income and of expense are the main types of financial records needed, although an inventory is necessary if financial data are going to be analyzed at the end of the year.

Enter all income from sheep—Farm records for the total farm program are most often used unless raising sheep is the only enterprise on a particular ranch. Under a diversified system of farming, it is essential to include all income from many enterprises although most producers prefer to enter income from each enterprise under a different column. In this fashion, it can be analyzed easily in terms of individual enterprises. Income includes all money or credit received from the sale of the main product (lamb, mutton, wool, feeders) and all by-products (breeding fees, etc.). Home consumed products and gifts of produce to others should also be entered, although on a large operation these items are often ignored. But they should not be overlooked with small farm flocks or youths who are fattening one or two head, as a misconception regarding profit is apt to arise if these gifts or non-cost items are forgotten.

Enter expenses immediately—For the most part, expenses will include all purchases (feed, minerals, fencing, etc.) and rent (pasture, leases, etc.). The important factor, however, is that every expense be written down immediately before it is forgotten. Minor items will mount up to a surprising total by the year's end if ignored or forgotten. Two main reasons for keeping track of all expense is that income taxes are apt to be lowered by such a record and cost of production figures will be incorrect unless every expense item is considered. More net profit is the ultimate result if this practice is followed. Expenses should include not only commonly thought of items, such as feed, labor and rent, but in addition, taxes, interest, repairs, fuel, minerals, water, breeding fees, new stock, veterinarian expense, legal fees, insurance, upkeep, and many others.

File all sale or expense slips—Many sheepmen carelessly throw away their invoices or other similar slips. Gasoline tax, for example, is a small part of each gasoline purchase, yet can add up to quite a figure in a year.

RECORDS FOR THE SHEEP BUSINESS

If all slips are kept, such things as not having to pay bills twice, proof for income tax purposes, etc., are guaranteed.

Prepare an inventory—An inventory is an itemized list of goods with their present estimated market values. The value represented by an inventory can, therefore, change even though the number of items remain the same. In addition, as items depreciate their value becomes less and less and this must be taken into account when preparing future inventories. Customarily, inventory is taken once a year, generally at the close of the calendar year or the fiscal year, although inventory could be taken at any slack season of activity. However, it should always be taken at the same time each year. In this manner, yearly comparisons of costs, profits, etc., can be correctly evaluated and serve to plot future progress.

An inventory for the sheep business will include most of the following:

1. A list of livestock—ewes, rams, feeders, etc.
2. Feed on hand—hay, grain, salt, minerals, etc.
3. Equipment—trucks, wagons, tractor, tools, etc.
4. Supplies—fuel, fencing, medicines, elastrator rings, etc.
5. Miscellaneous.

A major item such as land or buildings will depend on whether the operator is an owner or renter. As a renter, he would include the use of all buildings or pastures rented simply as an expense in the journal. In determining costs of production or other similar figures for sheep in a diversified program, it may be necessary to charge only a portion of the total value of land and buildings to the sheep enterprise in order to show a true cost. Caution should be used in assigning values to inventoried items, as the figures may become misleading if extremes are used. The best practice is to assign a conservative figure representing the average selling price at that date.

3. Keeping Production Records

Records take guess work out of the enterprise. Production records refer to those facts other than financial that have to do with the managing and producing side of the business. Number of lambs dropped, number of twins, yield of wool per ewe, or weight of lambs when marketed are the kinds of information needed to keep production records.

Select records in keeping with your farm—What kind of records to keep will depend to a large extent upon how much you wish to know about your business. Large commercial operators are generally satisfied to know total weight of wool and lamb produced plus the percentage of lamb crop; others, especially progressive breeders raising foundation stock, may wish to know such things as weight of fleece per ewe, number of lambs raised per individual ewe, weaning weight or marketed weight of each lamb, in order to intelligently plan a breeding program leading to production of superior animals. Purebred breeders will keep additional records such as those associated with the breed associations. It is a desirable practice to select the minimum number of records that best suit your conditions and then faithfully and accurately enter all pertinent information. (See Chapter III for purebred record card.)

Keep minimum records—Every successful breeder has his ideas as to the minimum number of records. However, a sound, basic list of production records would include:

1. Number of lambs dropped.
2. Number of lambs raised.
3. Number of ewes and their condition and age.

RECORDS FOR THE SHEEP BUSINESS

HERDSMAN POCKET FIELD BOOK

Date	Ewe No.	Sex	Birth Wt.	Notch	Tag	Remarks

Fig. 12.1—Any small notebook can be adopted, like the diagram above, to individual needs.

4. Pounds of wool produced (total or per animal).
5. Amount of feed consumed or condition and capacity of pastures.
6. Weight of feeders sold or fat lambs marketed.
7. Individual production records—include items as poor milker, refused to take lamb, etc.

A small vest pocket herdbook is desirable to keep on the sheepman's person so information can be jotted down while it is still accurate.

Maintain a filing system—Information from herd-books, shipping receipts, etc., should be entered on permanent record and account books. These records should then be filed under a simple filing system so **anyone** can read and evaluate them at a later date. All records should be maintained in safe, fireproof quarters. If a room or office or even a desk or part of a room can be set aside at home, records are much more apt to be accurately kept and used. Furthermore, buyers approve of businesslike transactions and surroundings.

BILL OF SALE
CATTLE, HORSES, SHEEP, (HIDES)_____ 19___
In consideration of_____dollars paid to me,
by_____of_____(State)
I hereby bargain and sell unto above named party_____head of_____
(hides) of which I am the owner. Animals, (hides) are now at_____
_____(Address), and described as follows:

No.	Horses Burros Mules	Steers Cows Calves	Sheep	Brand	Location of Brand	MARKS Right	MARKS Left	
						⌒	⌒	Seller_____ Address_____
						⌒	⌒	Witnesses _____
						⌒	⌒	Freeholders of
						⌒	⌒	_____ County,
						⌒	⌒	_____ (State), for two years last past.
						⌒	⌒	Subject to State and City Inspection. Seller agrees to reimburse buyer for losses incurred through inspection.

Fig. 12.2—A suggested bill of sale. The seller should provide a bill of sale, and the buyer should demand one. Bill of sale books may be purchased in supply houses.

Provide bill of sale—A bill of sale should be furnished with all animals sold and sheepmen should insist on receiving one when purchasing livestock. Simple forms may be purchased or prepared at home to include all necessary facts.

Purebred Records

Apply for registration papers immediately—Prompt, accurate recording, application, and follow-up of sales and association entries avoid delays and mixups and enhance the reputation and respect of a purebred breeder.

Double check entries—Most associations will say that approximately one-half of all entries received must be returned for further information. Therefore, double check to be certain that all information is correct. Look to see if:

1. Name and registration number agree with association records.
2. Duplicate names are used.
3. Name and sex agree.
4. Names are too long.

Examine tattoos—See that ear labels or tattoos are correct and agree with entry application.

Include all information—Many registration applications are sent with incomplete information. It is important, in addition, to be consistent in such things as always signing your name in the same fashion.

Follow through—Sales and transfer applications should be mailed to the proper parties as rapidly as possible. The very day or evening a sale is made is the best procedure. Delay here may discourage a customer from returning in the future.

Write it down—Do not rely on your memory for important facts such as date of transaction, breeding and lambing dates, weight of lambs or fleece. The more information one has, the more important it is to write it down and, consequently, the more reliable the records will be when used.

Keep informed on your flock—Sheepmen should at all times be able to answer, as a minimum, the following questions about their herd:

1. Number in the flock.
2. Number of ewes raising lambs.

Application for Registration

ENTRY BLANK—SOUTHDOWN SHEEP

Sire and Dam Recorded
_____

Post Office.. Date.................................., 19.........

Secretary, American Southdown Association, State College, Pa. Please record the following described animal:

Name of Animal to be Recorded.................................Date of Birth................., 19 ... Sex............ Ear Label.............

(Twinned with..)

If imported: When?.............................By whom?.............................P. O..................................

Bred by..P. O..................................

Owned by..P. O..................................

(Please ditto where owner and breeder remain the same)

Sold to..P. O..................................Date...........................

Name of Sire...Ass'n No..........Name of Dam..........................Ass'n No...........

The above is correct to the best of my knowledge and belief. { ....Breeder
....Owner }

Fig. 12.3—One kind of application blank. Purebred breeders should keep a supply of their breed blanks handy so application is not delayed.

3. Number culled.
4. Number left to breed. This is the inventory number.
5. Number of lambs weaned. (This generally represents the number of fat or feeder lambs sold.)

Tallies should be checked and a record kept. This tally is best made at shearing and weaning time for large herds.

4. Analyzing Records

Unless records are to be analyzed there is little value in keeping them. This applies especially to production records as financial records are often kept for legal reasons. Production records are just as important though, if analyzed, as they point the way to future operations. For example, a producer may feel he is raising a satisfactory number of lambs each year, yet a check may show his lambing percentage to be 110% while his neighbors are up to 125% or more. It is not impossible that such a difference could go unnoticed unless records were kept and analyzed. As a result, he could revamp his lambing and culling operations to bring his percentage up to a reasonable standard. The same situation is true regarding market weights, net profit, fleece weights, etc. While each producer may not keep the same kind of production records or wish to analyze them for the same purpose, the following results can be obtained by evaluating records:

1. Determine the percentage lamb crop.

$$\frac{\text{Number of lambs raised}}{\text{Number of breeding ewes}} \times 100 = \text{\_\_\_\_\% lamb crop.}$$

2. Determine cost of producing a pound of lamb.

$$\frac{\text{Total expenses minus wool value and by-products}}{\text{Total pounds of lamb sold}} = \text{\_\_\_\_cost per lb. of lamb.}$$

This method is a simple way, suitable for most producers.
3. Determine average weaning weight of lambs.
4. Determine average pounds of wool produced per ewe.
5. Determine the number of pounds of mutton or lamb produced per acre of irrigated or permanent pasture.
6. Determine the per cent of death loss.
7. Determine the average rate of daily gain when feeding out feeder lambs.
8. Determine the value of rent furnished by raising sheep.
9. Determine the value of beginning and closing inventories.
10. Determine average price received for breeding stock and per pound for fat animals.

There are other results possible to summarize from a well-kept set of records. No one sheepman will want to figure all of these, yet many have a direct bearing on the success of each producer.

Prepare a financial statement—Every sheepman should be able to prepare a financial statement. It is valuable if one plans to obtain credit, as all loaning agencies will demand one. In addition, a yearly financial statement is a guide as to how one is progressing in his total economic operations. A financial statement is a record of all assets minus all liabilities as of a certain date. **Assets** include such things as cash on hand, value of livestock, stocks and bonds, bills due you, land, buildings, equipment, feed and supplies, cash value of life insurance, or any other item that has worth regardless of whether or not it is totally owned by the person making the statement. **Liabilities** include all debts, bills owed, mortgages, loans, etc.

When **liabilities are subtracted** from **assets,** the figure is the **net worth** which will give the producer some idea of how he is progressing from year to year.

Make yearly comparisons—Too many producers fail to do this and yet it is the most advantageous way to evaluate the progress one is making. Comparisons can only be made if records are carefully analyzed.

5. Establishing Credit

A good credit record is one of the most desirable assets a rancher can possess. If one has a good credit rating, he is in a position to take advantage of market trends. In addition, should one have financial reverses or wish to expand, he will be able to follow the most advantageous direction with ease and ultimate profit. Good character, dependability and honesty are basic factors in establishing a good credit rating. However, these alone are not enough as loaning agencies and businesses also consider the time factor.

Establish credit early—It is surprising how long some people can live in a community, pay all bills promptly, and yet have a poor credit rating. Therefore, it is a desirable practice to establish credit by making **many** purchases at **one** good store, purchase **needed** article on time payments, use a monthly checking system to pay bills and all other methods that will entail some written record revealing how you spend your money. Frequently such materials as lime or fertilizer can be purchased on time and the increased yield more than pays all interest and other charges plus building up your soil and reputation. Many items can be purchased on time, interest free.

Do not abuse credit—This practice is dangerous and, of course, speaks for itself. Credit will not be misused if only those things which you have to purchase are bought on time, or if the article purchased will always be worth what you originally paid for it or more. For example, land and livestock are good credit risks; luxury items are not.

Use credit cards—Gasoline and other supplies may be purchased by obtaining a credit card. Such a card gives you a written record of all supplies purchased and builds up your rating with the company involved.

Pay bills promptly—Oftentimes penalties will be invoked if taxes or other bills are not satisfied on time. Discounts are often given for prompt payment. It may be necessary to borrow money to make correct payments. Occasionally, plain carelessness is responsible for bills not being paid when they are due. At any rate, it is extremely advisable that all bills be paid on or before the due date if a good credit rating is desired.

Buy advantageously—Proper use of credit will enable you to follow this practice if ready cash is not available. Off season buying of fertilizers, fuel, seed, and similar items will result in substantial savings as well as having needed items available when they should be. Even land and especially livestock have market trends that can be used to advantage when making purchases. It is desirable to make your dollar go as far as possible. Establishing credit early and using it wisely will result in greater net profit.

CHAPTER XIII

PRODUCING LAMB FOR THE HOME LOCKER

Good eating has long been associated with farm life. Fruit, vegetables, milk, butter, and most important of all, meat and delicious foods are common to rural living. With improved methods of curing and freezing, all perishable foods, and meat in particular, have taken on a new meaning in our diet. Lamb and mutton are high on the list of meat delicacies. Sheep products have an additional value to many people in that they can be easily raised on suburban or small, part-time farms. Even those who work in the city and spend most of their time away from home can supplement their income by raising a few head of sheep each year. Sheep do not have to be milked every day as do dairy cows, and they are quiet and seldom disturb neighbors. In addition, they require only a minimum of fencing and shelter. Weeds in fence rows, under trees in a family orchard, or other hard-to-get-at places can be kept down easily by raising a few head of sheep. Commercial producers, in order to supply their table, simply have to select the number desired for the year from their flock and slaughter them when ready.

Activities Which Involve Approved Practices

1. Determining the amount of lamb needed.
2. Determining the supply of feed.
3. Securing adequate refrigeration.
4. Utilizing waste feeds.
5. Purchasing feeders.
6. Using home-grown wool.
7. Raising the orphan lamb.

1. Determining the Amount of Lamb Needed

An 80-pound lamb produces a dressed carcass weighing about 36 to 40 pounds. Therefore, the entire carcass is small enough so that the average family refrigerator can hold the meat until it is consumed.

Eat more lamb—Each family is going to have to determine its own needs, as national averages are of little value at this point. About two-thirds of the lamb eaten in the United States is consumed in the section that lies north of Washington and east of Pittsburgh. Therefore, most families could use much more than they now do, especially in rural areas. As a minimum, those producing their own lamb should use two lambs per year. Many families will consume more, some up to 10 or 12 head. The amount to be used can be determined from these figures: According to **Farmers' Bulletin 1807,** the trimmed leg roasts from a 40-pound carcass will weigh about 6 pounds each and the shoulder roasts about 5 pounds each. There will be 7 pounds of breast and neck, and 8 pounds of loin and rib to be roasted or cut into about 30 medium-thick chops. These figures are based on 80 pound live-weight animals. Larger ones will give cuts proportionately higher, as many people prefer well-fed lambs weighing up to 100 pounds or more.

2. Determining the Supply of Feed

Whether or not a person will raise sheep or any livestock depends upon the availability of feed.

Find your feed first—It may not be profitable to raise sheep on many farms regardless of size if feed is not available. Many people commit the error of getting their livestock before the feed question is settled. Therefore, it is essential to first determine whether an adequate supply of feed is available.

Determine the amount of feed—As a rough guide, five sheep will equal one animal unit or eat as much as does one mature cow. **USDA Bulletin 1753** says that the average cow will eat three to three and one-half tons of hay and one of concentrates per year. Therefore, this same amount would feed five head of mature ewes. If lambs were slaughtered right off their mothers as fat lambs, they could probably be added to this estimate with little increase in feed requirements. However, it is somewhat difficult to determine accurately sheep needs if only a few head are raised, as cleanup of fence rows, leaves, lawn clippings, etc., may form a sizable portion of the diet. People who raise only three or four head for home consumption are apt to get the idea that sheep eat nothing at all because of the cleanup of waste around the farmstead. At any rate, it is desirable to find out how much feed will be available and then get enough sheep to properly utilize it, because in most instances, if all feed must be purchased, it is not wise to try to raise your own meat.

Utilize irrigated and permanent pastures—It will pay most farmers on small farms to plant pasture mixes and raise their own feed, especially if irrigation water is available. As a rule, one to two acres would supply enough feed for five to ten head of sheep year round, depending on locality. Extra feed may be cut for winter hay. In some states, grazing may be year round if the right grasses and legumes are used.

3. Securing Adequate Refrigeration

New, improved mechanical refrigeration has made tremendous strides in keeping meat fresh on the farm. In fact, the diet of many farm families has been considerably changed as a result of the development of the home freezer.

Obtain adequate size—As a general rule, it is better to get a freezer of larger capacity than the calculated size needed. Larger freezers do not have to overwork to keep meats safe and are able to handle excess emergencies easily. The total cost of a larger box is more than a small one, but much cheaper when figured on a cubic foot basis. Manufacturers can supply information regarding the proper size, but most farm families will find adequate storage with a freezer of 20 cubic foot capacity.

There are few advantages and some disadvantages for a walk-in type box. Walk-in boxes should be used only if the storage capacity required is very great.

4. Utilizing Waste Feeds

Here is one of the real advantages and an impetus for many farms, large or small, to produce a few head of sheep even if only enough to supply the home locker. The amount of waste feed on most ranches is often surprising. In addition, sheep are short enough so they seldom damage trees or shrubs yet will get into out-of-the-way places to give a "cleanup" effect to the farmstead, ditchbanks, or barnyard.

Feed only unspoiled products—With many waste feeds, particularly by-products, spoilage is always a danger item. Even if sheep are not killed as a result of spoiled feed, they may go off feed for many days and lose weight. Watch out particularly for musty, damp feeds. Therefore, it is desirable to feed only fresh waste products or those properly cured.

Investigate all foodstuffs—Free feed results in cheap meat, therefore, no source of feed should be missed. Almond hulls for example, are good sheep feed, especially for over-wintering ewes. This product, only recently wasted, is now eagerly sought by sheepmen. Lawn clippings, waste materials (not garbage) from stores and markets, as well as excess vegetable or crop products from the farm can be converted into lamb and mutton. Weeds along fence rows and ditches or lanes and under trees will be eaten. Some sheepmen turn their ewes into vineyards to clean up the leaves on the vines with favorable results. Bean straw, stubble, and many other farm by-products are good sheep feeds, especially when supplemented by a protein concentrate.

5. Purchasing Feeders

Many persons who wish to produce their own lamb or mutton will not want to bother with ewes on their place the year round. Under these conditions, the growing out of several feeders will satisfy their needs.

Purchase disease-free stock—Sometimes animals may be offered for sale because they are ailing. Weak or sick lambs seldom, if ever, make a profit and are a continual source of trouble. Select alert, parasite-free feeders for fattening. If you know their origin, better yet, as you know what to expect in the way of parasites or disease.

Do not purchase fat lambs—Lambs in extra good condition generally go down hill, especially when their feed is changed and when they are put on poorer quality feed. Thin, potbellied animals may indicate sickness; therefore, that in-between quality exhibited by good feeders is the type that should be selected.

Select growing stock—Generally speaking, this means young, healthy animals but in only fair condition. Lambs that are weaned, but not milk fat, make top-notch feeders to grow out for the home locker.

6. Using Home-Grown Wool

Wool fiber, as everyone knows, makes excellent material for cloths, blankets, and a host of other uses. Wool material is also expensive when purchased retail. Many growers with only the wool of a few head of sheep either cannot command the attention of a buyer or cannot find a nearby market for their wool. Under these conditions, a desirable practice may be to utilize your own wool for goods needed by the farm family.

Have your wool processed—There are numerous companies that specialize in processing small lots of wool into usable woolen articles. These goods are warm, serviceable, neat appearing, and compared with retail prices, can be obtained at a considerable saving. A sense of pride in producing and wearing your own apparel is an added benefit.

Two suggested companies that will process your wool are:

1. Baron Woolen Mills, Brigham City, Utah   84302.
2. Terrace Yarns, 4038 S.W. Garden Home Rd., Portland, Oregon   97219.

Wool should be as clean as possible, not scoured, and sent by express or parcel post in suitable containers to the company of your choice. Information as to prices, amount of wool to ship, and how to ship may be obtained free by writing to any of the woolen companies.

7. Raising the Orphan Lamb

On a small farm an economical source of meat may be the orphan lamb. Large growers sometimes dislike to bother with orphan or disowned lambs and will either give them away or sell them cheaply to anyone who happens to be around. Even in large flocks, raising an orphan lamb makes a good project for the farm boy or girl. In any event, if surplus labor is available, it is a good idea to raise orphan lambs as they make an excellent source of meat for home consumption.

Get another ewe to adopt, if possible—There is no substitute for milk when it comes to raising a lamb both from the standpoint of nutrition and labor involved. **Iowa Bulletin P107** lists several ways to get a ewe to adopt a strange lamb:

1. Sprinkle some of her milk on the lamb's rump and on her nose.
2. Tie ewe with a small halter in lambing pen with orphan lamb.
3. Be patient—try holding ewe the first few times lamb nurses.
4. Skin ewe's dead lamb and tie skin onto orphan.

Raise orphans by hand—Get milk from a fresh ewe, if possible, especially for the first few days. Use an ordinary baby bottle and nipple. After several days, cows' whole milk may be used. It is important to feed **small** quantities **frequently** if the digestive system is to be kept functioning properly, such as an ounce or two every two hours at first. Then gradually change this schedule until three feedings per day are fed after ten days. Keep the milk warm and nipples and bottles clean. Induce the orphan to eat grain and pasture as soon as possible. The length of time milk is fed will vary with the availability of labor, milk, and the speed of growth desired for a particular lamb. Although milk

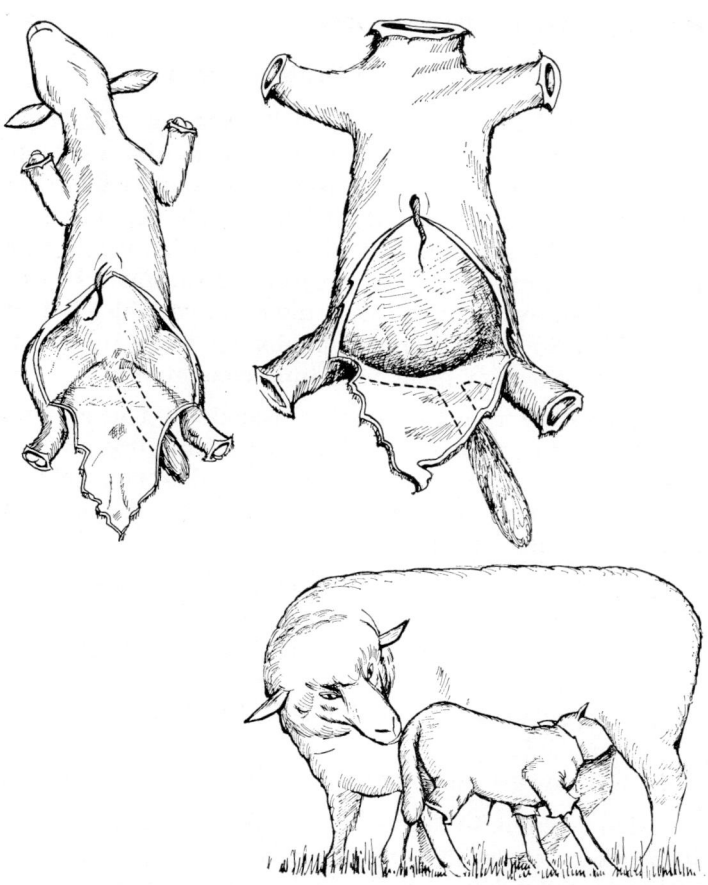

Fig. 13.1—Top left: Hide grafting is best accomplished by skinning in the manner illustrated so that a triangular flap of skin is loosened to hang down at the rear of the grafted orphan. After cutting off the fore and rear legs at the knee and hock joints, the hind leg may be doubled at the stifle joint so that the cuff shown on the hock may be slipped over the end of the leg. The tail can be severed at the base and the hide slipped forward over the head to come off fleshside out as with a person removing a slip-over sweater.

Top right: As the second step, the skin is severed from around the neck of the dead lamb and again turned fleshside in. It is now ready to go on the orphan.

Bottom: Note that the rear of the orphan is totally covered by the pelt of the dead lamb, the principal opening of the skin being hidden below. Now the mother of the dead lamb, searching for identification by odor, detects only that of her own young. The orphan should be tied down in the same location where the dead lamb was found for 5 or 10 minutes after being dressed in the skin. This will allow the new mother to become acquainted with the lamb and guard against an unnatural display of activity.

Adapted from **California Agricultural Extension Service Manual 40, University of California**)

can be discontinued after several months without endangering the health of the lamb, this will cause a decrease in rate of gain.

The following information is provided by **University of California Agricultural Extension Service Manual 40:**

One of the commonest mistakes made in rearing orphan lambs is overfeeding, which may result in the animal's death. Keep young lambs hungry for the first few days. Although it is difficult to suggest definite amounts of milk because of variation in size and vigor, the following schedule is generally satisfactory.

The first two days—feed 2 to 3 ounces at least four times a day.

The second two days—increase the feeding by 1 or 2 ounces and feed four times a day.

The next week—feed 4 to 6 ounces four times a day.

The week after that—feed 6 to 8 ounces four times a day.

Then gradually change the lamb over to three feedings a day of one pint per feeding. As it grows older, it may safely take two quarts a day given in two or three feedings. Lambs will vary in their milk consumption, and the attendant must be the judge.

After six weeks, skim milk, if necessary, may be substituted for whole milk, but the lambs will gain better on whole milk. They should be fed milk until they are at least three months old.

Lambs will soon begin to eat hay and grain (at two or three weeks of age). A small amount of bright alfalfa hay should be kept before them at all times. The grain mixture may consist of any of the common farm grains. Calf meals are often used with good results.

Give the lambs free access to pasture at an early age.

CHAPTER XIV

ESSENTIAL SKILLS FOR THE SHEEPMAN

In order for the sheep producer to be able to raise, market, and care for his livestock, he should be able to perform reasonably well a high percentage of the everyday skills necessary to tend sheep properly. The following list should prove of value, especially to begin-

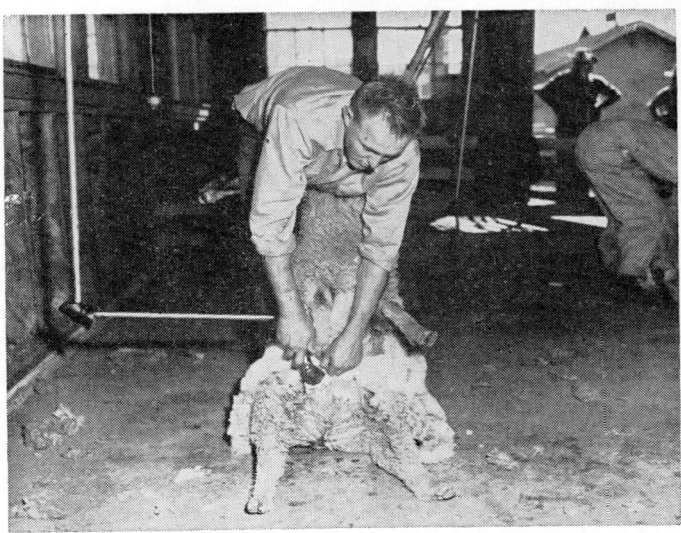

Fig. 14.1—Knowledge of how to properly tag and shear sheep is a valuable skill for any sheepman to possess.

Fig. 14.2—Two approved practices that sheepmen must learn and practice for success in the sheep business are well depicted by these illustrations. Top figure shows the proper way to grasp a sheep prior to setting up for tagging or examination; lower photo shows the correct method of tagging a sheep so vent and udder are free of long wool.

ners, as the majority of these skills are manipulative and involve **doing** ability. Therefore, every effort should be made to gain experience **before** investing too heavily in the sheep enterprise.

The first group of skills are those which are most essential for the average sheepman who is producing in a commercial fashion.

A. Essential Skills

1. Catch and hold sheep properly.
2. Tag ewes and assist at lambing time.
3. Disinfect navel of lamb.
4. Give lamb to foster mother.
5. Dock and castrate lambs.
6. Ear notch and ear tag.
7. Trim feet ("throw" sheep and set up ram).

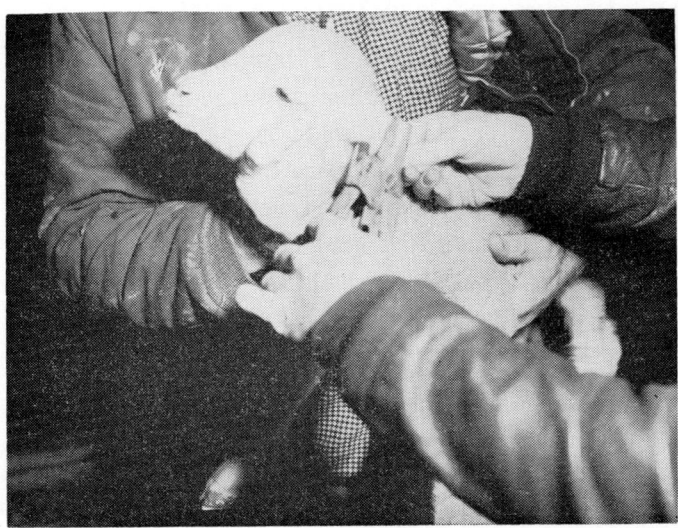

Fig. 14.3—Putting a metal identification tag in the ear of a lamb. An easy skill to learn.

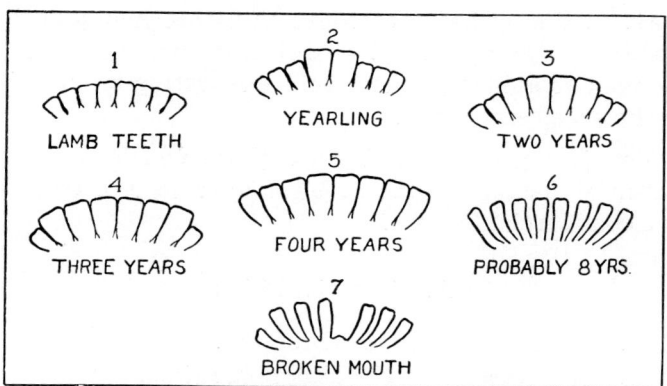

Fig. 14.4—This diagram illustrates how easy it is to learn to tell the age of sheep by their teeth. Permanent teeth are not only larger, but whiter than lamb's teeth.

8. Recognize and treat common ailments (wool maggots, screw worms, foxtail infection, founder and bloat, stomach worms, ticks, lice).
9. Treat to prevent shipping fever.
10. Manage creep feeding of lambs.
11. Select wethers and lambs for feeding.
12. Move ewes and young lambs.
13. Brand sheep and paint a ram.
14. Load and haul lambs in trucks.
15. Tie a fleece and pack a bag of wool.
16. Dodge sheep through chute.
17. Determine age of sheep.
18. Sharpen sheep shears.
19. Use a lamb jail.

There are many other skills involved and as one gains experience or begins to specialize in sheep or in some particular phase, such as the purebred industry, other skills and abilities will become important and should be practiced.

B. Additional Desirable Skills

1. Skin a dead lamb.
2. Butcher a sheep.
3. Keep flock records.
4. Determine grade and value of wool.
5. Shear sheep.
6. Determine shrinkage of fleece.
7. Select rams and ewes.
8. Grade market lambs.
9. Make a gunny sack blanket for sheep.
10. Block and train for show.
11. Sew or clip "in-turned" eyelids.
12. Make a sheep halter.
13. Pull teeth from aged or broken-mouth ewe to make the ewe a "gummer."

APPENDIX A—Associations and Publications

Sheep Breed Associations

American Cheviot Sheep Society, S. R. Gates, Secy., Lafayette Hill, Pennsylvania 19444
American Corriedale Assn., Inc., Russell Jackson, Secy., P. O. Box 29C, Seneca, Illinois 61360
American Cotswold Record Assn., Virgil A. Bortel, Secy., 11903 Milwaukee Rd., Britton, Michigan 49229
American & Delaine-Merino Assn., Harold E. Simms, Secy., Aleppo, Pennsylvania 15310
American Hampshire Sheep Assn., Roy A. Gilman, Secy., Stuart, Iowa 50250.
American Oxford Down Record Assn., James Hanson, Box 401, Burlington, Wisconsin 53105
American Rambouillet Breeders Assn., 2709 Sherwood Way, San Angelo, Texas 76901
American Romney Breeders' Assn., John H. Landers, Jr., Room 212, Withycombe Hall, Oregon State College, Corvallis, Oregon 97331
American Shropshire Registry Assn., Inc., Mrs. Elizabeth Glasgow, Secy., P. O. Box 1970, Monticello, Illinois 61856
American Southdown Breeders Assn., W. L. Henning, Secy., 212 South Allen St., State College, Pennsylvania 16801
American Suffolk Sheep Society, Allen Jenkins, Secy., 52 N. 1st East, Logan, Utah 84321
Columbia Sheep Breeders Assn., Richard L. Gerber, Secy., P. O. Box 272, Upper Sandusky, Ohio 43351

Continental Dorset Club, J. R. Henderson, Secy., P. O. Box 1206, Carbondale, Illinois 62901

Karakul Fur Sheep Registry, Annette S. Harris, Secy., Fabius, New York 13063

Montadale Sheep Breeders Assn., Anne Gregory, Secy., P. O. Box 1007, E. St. Louis, Illinois 62204

National Lincoln Sheep Breeders' Assn., Ralph O. Shaffer, Secy., 5284 S. Albaugh Rd., West Milton, Ohio 45383

National Suffolk Sheep Assn., Mrs. Betty Biellier, Secy., P. O. Box 324S, Columbia, Missouri 65201

National Tunis Sheep Reg., Ralph E. Owen, Secy., Fulton, New York 13069

Texas Delaine-Merino Record Assn., G. H. Johanson, Secy., Rte. 1, Brady, Texas 76825

U. S. Targhee Sheep Assn., Gene Coombs, Secy., Box 2513, Billings, Montana 59103

Other Sheep Associations

American Lamb Council, Dept. LC-171, 200 Clayton St., Denver, Colorado 80206

California Wool Grower's Assn., 3382 El Camino Ave., No. 6, Sacramento, California 95821

Idaho Wool Growers Assn., M. C. Claar, Secy., 17 Broadbent Building, Box 2596, Boise, Idaho 83702

Midwest Wool Marketing Cooperative, Inc., 911-17 Wyoming St., Kansas City, Missouri 64101

Montana Wool Growers Assn., Livestock Bldg., Helena, Montana 59601

National Wool Growers Assn., 600 Crandall Building, Salt Lake City, Utah 84101

North Dakota Wool Growers, Fargo, North Dakota 58102

Wyoming Wool Growers Assn., J. B. Wilson, Secy., McKinley, Wyoming 82226

Goat Breed Associations

American Angora Goat Breeders' Assn., Mrs. Thomas L. Taylor, Secy., Rocksprings, Texas 78880

American Dairy Goat Assn., Don Wilson, Secy., P. O. Box 186, Spindale, N.C. 28160

American Goat Society, Inc., J. Willet Taylor, Secy.-Treas., 1606 Colorado, Manhattan, Kansas 66502

Sheep and Goat Publications

Better Goat Keeping, Ipswich, Massachusetts 01938
Dairy Goat Journal, Columbia, Missouri 65201
Sheep and Goat Raiser, Cactus Hotel Building, P. O. Box 189, San Angelo, Texas 76901
The Bleat, Box 25, Victoria, British Columbia
The Goat World, Roanoke, Virginia 24000

Organizations Allied with Livestock Production

American Farm Bureau Federation, William J. Kuhfuss, Pres., 1000 Merchandise Mart, Chicago, Illinois 60654; Roger Fleming, Secy.-Treas. & Dir., Washington Office, 425 13th St., N.W., Washington, D.C. 20004

American Meat Institute, 59 East Van Buren St., Chicago, Illinois 60605

Livestock Conservation, Inc., Executive Director, 405 Exchange Building, Chicago, Illinois 60609

National Association of Animal Breeders, Dr. H. A. Herman, Exec. Secy., George Nichols, Asst. Secy., 512 Cherry St., P. O. Box 1033, Columbia, Missouri 65201

National Farmers' Union, Tony T. Dechant, Pres., Edwin Christianson, Vice Pres., Kenneth L. Motz, Secy.-Treas., Box 2251, Denver, Colorado 80201

The National Grange, John W. Scott, Master, Robert G. Proctor, Secy., 1616 H. St., N.W., Washington, D.C. 20006

National Live Stock and Meat Board, David H. Stroud, Pres., 26 South Wabash Ave., Chicago, Illinois 60603

National Society of Live Stock Record Assoc., Don Jones, Pres., Allan C. Atlason, Secy.-Treas., 3964 Grand Ave., Gurnee, Illinois 60031

APPENDIX B—Tables of Weights and Measures

Sheepmen frequently have occasion to measure, or to convert one measure to another. Measuring should be carefully and accurately done.

Liquid Measure

```
1 tablespoon ........................... 3 tsp.
1 ounce ............................... 2 tbsp.
1 cup ................................... 8 oz.
1 pint ................................. 2 cups
1 quart ................................. 2 pt.
1 gallon ................................ 4 qt.
1 barrel ........................... 31.53 gal.
1 cubic foot .................. 7.48 gal. water
231 cubic inches .................... U. S. gal.
1 liter ...................... 61.023378 cu. in.
                   ............... 0.035314 cu. ft.
                  ............ 0.264170 U. S. gal.
                   ............ 0.2201 imperial gal.
1 U.S. quart ....................... 0.946 lit
1 U.S. gallon ...................... 3.785 lit
```

Dry Measure

```
1 quart ................................. 2 pt.
1 peck .................................. 8 qt.
1 bushel ................................ 4 pk.
1.244 cubic feet ......................... 1 bu.
```

Weight

1 pound	16 oz.
1 ton	2000 lbs.
1 long ton	2240 lbs.
1 gram	15.43 grains
	0.0353 avoir. oz.
1 kilogram	2.205 avoir. lbs.
1 metric ton	0.984 gross or long ton
	1.102 net or short tons
1 avoirdupois ounce	28.35 g
1 avoirdupois pound	0.4536 kg
1 ounce (liquid)	29.573 cc
1 cubic centimeter	0.034 oz. (fluid)

Water Equivalents

1 gallon	231 cu. in.
1 cubic foot	7.5 gal.
	62.4 lbs.
1 acre foot	43,560 cu. ft.
1 acre inch	3,631 cu. ft.
	27,154 gal.
1 acre inch per hour	450 gal./min.
1 cubic foot per second	450 gal./min.
1 lb./sq. in. pressure	2.31 ft. head
1 foot head	0.433 lb./sq. in.

Feed Weights

12 ears of corn	7.5 lbs.
1 gallon shelled corn	7 lbs.
1 gallon ground corn	6 lbs.
1 gallon of corn & cob meal	5.5 lbs.
1 gallon wheat or kafir	7.5 lbs.
1 gallon ground wheat	7 lbs.
1 gallon oats	4 lbs.
1 gallon ground oats	3 lbs.
1 gallon barley	6 lbs.
1 gallon ground barley	4.5 lbs.
1 gallon molasses	12 lbs.
1 gallon bran	2 lbs.
1 gallon shorts	3.5 lbs.
1 gallon linseed or cottonseed meal	5 lbs.
1 gallon tankage	6.5 lbs.
1 gallon alfalfa meal	2.5 lbs.
1 gallon milk	8.6 lbs.
1 cubic foot hay, loose	4 lbs.
1 cubic foot hay, baled	12 lbs.
1 cubic foot hay, chopped	10 lbs.
1 cubic foot straw, loose	3.5 lbs.
1 cubic foot straw, baled	9.5 lbs.

TABLES OF WEIGHTS AND MEASURES

1 cubic foot corn, ear .... 28 lbs.
1 cubic foot corn, shelled .... 47 lbs.
1 cubic foot wheat .... 48 lbs.
1 cubic foot oats .... 27 lbs.
1 cubic foot barley .... 40 lbs.

Square or Surface Measure

144 square inches .... 1 sq. ft.
9 square feet .... 1 sq. yd.
160 square rods .... 1 a.
43,560 square feet .... 1 a.
640 acres .... 1 sq. mi.
1 square mile .... 1 section
36 sections .... 1 township
23,040 acres .... 1 township
1 square centimeter .... 0.155 sq. in.
1 square meter .... 10.76 sq. ft.
.... 1.196 sq. yds.

See **Handbook of Livestock Equipment**, The Interstate Printers & Publishers, Inc., Danville, Illinois, for complete tables of standards for livestock industry, including sheep.

APPENDIX C—Summary of Approved Practices in the Sheep Enterprise

CHAPTER I

Opportunities in the Sheep Industry

1. Success in the sheep business is possible only if you can grow grass or produce feed suitable for sheep economically on the land under your control.
2. It is essential to learn the facts necessary to the proper handling of sheep.
3. The greatest increase in sheep numbers will come about by increasing the number of farm flocks.
4. Those interested in becoming purebred breeders should first consider developing a successful commercial herd.
5. One method of entering the sheep business on relatively small acreages of land is by utilizing irrigated pastures.
6. Before entering the sheep business one must be sure of the availability of an adequate source of economical feed.
7. There is a trend toward intensive type lamb production, which includes ewes producing two or more lambs with each pregnancy, improved diets, faster gaining lambs, and year-round production.

CHAPTER II

Selecting the Breeding Stock

1. Select a type and breed adaptable to your farm and market conditions.
2. Buy thrifty lambs weighing from 55 to 65 pounds and about four months of age if they are to be used for fattening.
3. Large two-year-old ewes with sound teats and udders make good breeding ewes.
4. Yearling rams, healthy and capable of improving the quality of ewes in your particular flock, are the best breeding rams.
5. For best buys, study the market and pay no more than market price.
6. Late summer or early fall is generally the most favorable time to purchase grade ewes.
7. Cull all ewes that do not meet the standard of your flock or produce a lamb.
8. Start with fewer sheep than farm can carry and grow into proper number.

CHAPTER III

Breeding and Improving Sheep

1. Select a sound method of breeding and stay with it.
2. Maintain a high percentage lamb crop by flushing ewes, conditioning rams, and using correct number of ewes per ram.
3. In general, breed ewes to lamb at two years of age.
4. Well-grown ewe lambs or large breeds such as Hampshires may be bred for the first time as lambs.
5. Cull year around and systematically at shearing time.
6. Keep 18 or 20 of the best ewe lambs for replacements for every 100 ewes.

SUMMARY OF APPROVED PRACTICES

7. Maintain the ram in medium flesh and provide exercise for best results.
8. Accurate records fairly used are the surest way to attain reasonable standards.
9. Follow a regular pattern of approved practices for improving sheep, preventing disease, and providing an adequate diet.

CHAPTER IV

Handling Sheep and Lambs

1. Shear sheep by machine.
2. Plan convenient shearing sheds with hardwood or canvas floors for proper shearing quarters.
3. Remove fleece in one piece and tie properly.
4. Use the sides of the palms to examine wool. Never grab wool to hold sheep.
5. Provide ample exercise and keep away from mud during winter.
6. Animals should go into the winter carrying good flesh. Emergency feed should be kept on hand.
7. Catch sheep by confining flock first and then catch individual sheep by rear flank.
8. Dock and castrate lambs from one to three weeks of age during warm weather on clean ground.
9. Tag sheep prior to shearing and ewes prior to lambing.
10. Trim hoofs regularly of breeding stock especially those on irrigated pastures.
11. Brand as small an area as possible using specially prepared branding fluids.
12. Learn to tell the age of sheep by examining their teeth.
13. Provide extra protection for show sheep. Train them to stand.

CHAPTER V

Raising Lambs

1. Keep bred ewes away from other livestock.
2. Feed bred ewes well so they do not lose flesh during pregnancy.
3. Separate heavier ewes from late lambers. Watch first lambers.
4. Be on hand during parturition, but assist only when necessary.
5. Make certain lamb is breathing. Disinfect navel. Assist weak lambs to nurse; warm chilled lambs.
6. Examine udders frequently, and milk out the udder if necessary.
7. Keep ewes in good flesh. Feed lambs fresh pasture.
8. Wean at three and one-half to five months of age.
9. Control predators, especially dogs.

CHAPTER VI

Feeding and Fattening Sheep

1. Feed your pastures a proper plant food.
2. Plant the right pasture mixture for your locality.
3. Supplement feed of sheep on poor pastures.
4. Get lambs to market weights quickly.
5. Good lambs should gain one-third pound per head daily.
6. Always supply salt, but feed other minerals according to your area.
7. Sheep require clean water free of parasites.
8. Do not feed sheep antibiotics.
9. Make good feed purchases by making off-season purchases in large quantities and buying direct from producers.
10. Take full advantage of the grazing habits of sheep

by feeding all possible by-products and waste feeds to which you have access.
11. Use pellets whenever competitive with other feeds.

CHAPTER VII

Shelter and Equipment for Sheep

1. Locate buildings correctly so they will be dry, sheltered, and convenient to pastures and to farm house.
2. Plan for minimum waste space and ease of cleaning.
3. Purchase sheet metal equipment. Use good grade, smooth lumber for feeders and racks.
4. Develop natural water. Build for ease of cleaning and use automatic water valves.
5. Use woven wire and preservative-treated fence posts.
6. Build cutting chutes and corrals from proper plans.
7. Keep a medicine chest on every sheep farm.
8. Provide correct loading and shipping facilities.
9. Locate storage near all-year road.
10. Build good fences using steel posts or treated posts.

CHAPTER VIII

Controlling Parasites and Diseases

1. Maintain healthy livestock by rotating pastures, eliminating mechanical hazards, and using proper sanitary measures.
2. Control dogs and predators by proper use of fencing and cooperation with government authorities.
3. Prevent bloat by providing grass legume pastures, and always providing some coarse roughage as well.
4. Use modern worm killers to control internal parasites.
5. Prevent wounds and tag properly to prevent fly strike.

6. Control shipping fever by eliminating hardship conditions.
7. Reduce losses from poisonous plants by eradicating toxic species and fencing off or draining danger zones. Herd animals slowly when moving through hazard area.
8. Separate sick animals until cause of illness is determined.
9. Vaccinate for contagious diseases only when recommended by veterinarian.

CHAPTER IX

Butchering Lamb and Mutton on the Farm

1. Thrifty, well-finished lambs weighing 80 to 100 pounds make the most desirable carcasses.
2. Wash thoroughly and chill carcass for about 24 hours.
3. Cut up carcass into the recognized wholesale cuts.

CHAPTER X

Selecting and Using Lamb and Mutton

1. If possible, select lamb from mutton breeds.
2. Learn the difference between lamb and mutton and how to identify cuts of meats.
3. Package meat according to family size, label correctly, and freeze immediately.
4. Cook promptly after thawing.

CHAPTER XI

Marketing Mutton, Lamb, and Wool

1. Keep posted on prices and deliver only to known buyers.

SUMMARY OF APPROVED PRACTICES

2. Sort lambs and market at peak condition.
3. Handle wool as though it were perishable.
4. Investigate all selling channels.
5. Truck from range to rail head.
6. Rest before loading and load carefully.
7. Load proper number per car and partition to prevent piling up.
8. Start fitting young sheep eight to twelve weeks prior to show and aged sheep, four to six weeks.
9. Plan sales months in advance. Get the best auctioneer and assistants.
10. Complete transfers rapidly. Guarantee your animals.
11. Follow suggested rules in photographing sheep and discard all poor pictures.

CHAPTER XII

Records for the Sheep Business

1. Keep records simple and in such a form that they can be understood by anyone qualified to interpret livestock records.
2. Analyze all records in order to plan your future program.
3. Establish credit early but do not abuse it.
4. Pay bills promptly.
5. Buy advantageously by use of credit.

CHAPTER XIII

Producing Lamb for the Home Locker

1. Two 80-pound lambs per year will supply the average farm family.
2. Determine your supply of feed first before raising lamb for home consumption.

3. Obtain a freezer slightly larger than calculated needs.
4. Purchase only disease-free feeders to be grown out for home use.
5. Have your wool processed into usable woolen articles.
6. Raise orphan lambs by hand as a source of meat.

CHAPTER XIV

Essential Skills for the Sheepman

1. Learn to do the essential skills quickly and easily.
2. Obtain and keep on hand all necessary tools and equipment in order to perform essential skills.

APPENDIX D—Glossary of Common Terms

Antibiotic—Chemical compounds generally produced by molds that have the ability to inhibit growth of certain bacteria.

Braid—Seventh and coarsest of the U. S. grades of wool under the old system of naming the grades.

Brand—In sheep, refers to the markings made on wool with special branding fluids for identification purposes.

Break—Weak place in a fleece or staple of wool caused by malnutrition, overfeeding, or fever.

Breed—A group of animals possessing well-defined, distinguishing characteristics and which are able to reproduce these characteristics in their offspring with a reasonable degree of certainty.

Britch wool—From the lower parts of the thighs; often coarse and hairy.

Broken mouth—Sheep who have lost some, but not all, of their teeth.

Buck—Male sheep; generally refers to those of breeding age.

Carding—A manufacturing process that converts loose, scoured wool into a continuous strand suitable for subsequent operations.

Castrate—Removal of the testicles or reproductive organs from a male animal.

Character—Crimp, handling qualities and general appearance of wool.

Claiming pen—See jail.

Clip—The aggregate of all fleeces from a flock.

Colostrum—The thick, viscous milk produced by the ewe during the first week or so after lambing.

Combing wool—Long enough to comb on the English or Noble comb.

Condition

  Animal—The amount of fat on an animal. Fat animals are said to be well conditioned.

  Wool—The amount of yolk or dirt in grease wool. Wool is heavily conditioned if it contains large amounts.

Conformation—See diagram of sheep parts named. Refers to the shape and design.

Creep fed—Refers to animals (generally lambs) that are given extra feed by means of small openings in panels that permit the smaller animals to enter.

Crimp—Natural waviness of the wool fiber.

Crossbred—Offspring resulting from mating two pure breeds. (Wool from a crossbred sheep.)

GLOSSARY OF COMMON TERMS

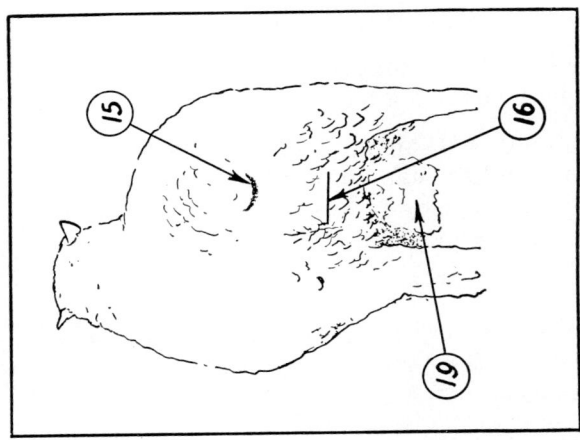

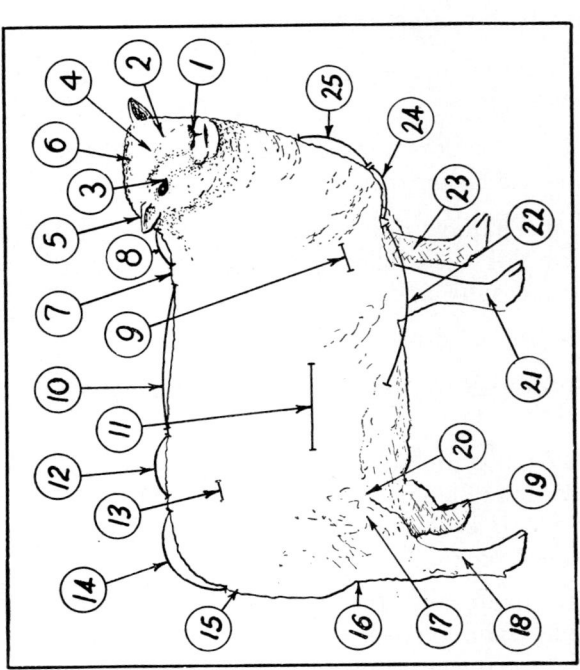

Fig. G.1—The external parts of a sheep: 1, Muzzle broad, lips thin, nostrils large; 2, face short, features clean-cut; 3, eyes large and clear; 4, forehead broad; 5, ears alert and not coarse; 6, poll wide; 7, top of shoulder compact; 8, neck short, thick, blending smoothly with shoulder; 9, shoulder thickly covered with flesh; 10, back broad, straight, thickly and evenly covered; 11, ribs long, well sprung, and thickly covered; 12, loins broad, thick, and well covered; 13, hips wide and smooth; 14, rump long, level, and wide to dock; 15, dock thick; 16, twist deep and firm; 17, thighs full, deep, and wide; 18, legs straight, short, and bone smooth; 19, cod or purse in wether, scrotum in ram, udder in ewe; 20, flank full and deep; 21, forelegs straight, short, and strong; 22, chest deep, wide and full; 23, forelegs wide apart and forearm strong; 24, brisket full and rounding in outline; 25, breast well extended.

Culling—Cull—Animal taken out of the flock because it is below herd standards.

Density—Refers to the number of wool fibers per square inch of surface area of skin.

Disease (See Chapter VIII on parasites and diseases)—Any condition other than normal health. However, it is often considered by farmers to mean an illness caused by microorganisms.

Dock—Removal of tail of sheep.

Dodge—Refers to the act of separating various kinds of sheep by means of some mechanical separating device such as a dodge gate.

Dress out—Removing the offal from sheep so that the carcass becomes an edible product.

Early lambs—Lambs born early enough to go to market as spring lambs.

Earmark—Slits or perforations made in an animal's ears for identification purposes.

Elastrator—A mechanical device used to apply elastic bands to the tail or testicles of sheep so that these tissues will atrophy and disappear.

Ewe—Female sheep of breeding age.

Exotic—The term used to describe animals foreign to a region.

Feeder lambs—Young animals under one year of age that carry insufficient finish for slaughter purposes but which show indications of making good gains if placed on feed.

"Fell"—A thin membrane found between the skin and carcass. Upon exposure to air, it hardens and protects the meat from drying out unduly.

GLOSSARY OF COMMON TERMS

Felting—The process of wool fibers locking together to form a mat.

Fill—The amount of feed and water in an animal.

Fine—First of U. S. market grades. Also wool of small diameter.

Finish—Refers to fatness; highly finished means very fat.

Fleece—Consists of the entire coat of wool as it comes from the sheep or while still on the live animal.

Flock (or band)—Refers to the total number of sheep under one management. Flock is often used in connection with small numbers on the farm, whereas band is used to designate large numbers on the range.

Flushing—The practice of feeding thin ewes more during the period two or three weeks immediately prior to breeding. Rations are generally high in protein content.

Foetus—Unborn young; refers to the young of an animal while it is still carried in the mother's uterus.

Foxhead—A device used to hold in an expelled uterus after it has been cleaned and replaced in the ewe. It is made of one-fourth inch wire in the shape of a foxhead.

Grade—Average diameter expressed in arbitrary terms of the fibers in a fleece.

Grease wool—Wool as it comes from the sheep. It contains large amounts of grease or lanolin.

Gregarious—The flocking instinct; tendency of sheep to bunch together.

Gummer—A sheep with all the incisor teeth missing.

Half-blood, ⅜ blood, etc.—Names of official grades of wool.

Hank—Unit of measurement of yarn in the wool textile industry; in worsted yarns on which the spinning counts are made and hence the names of grades; a hank is 560 yards of worsted yarn.

Heavy ewes—Female sheep indicating approaching parturition by their full sides and bellies and looseness around genital parts.

Hothouse lambs—Lambs born in fall or early winter and marketed when from 9 to 16 weeks of age or from Christmas to May to a special trade.

Jail—A small pen only large enough to hold one ewe and her offspring.

Kemp—A white, opaque, weak, and brittle fiber found in some fleeces of wool and mohair. It does not take dyes as wool does and has little value in manufacturing.

Lamb—A sheep under one year of age (usually under 14 months of age). If a lamb has lost any of its temporary teeth, it would not be classed as a lamb.

Lambing—Giving birth to lambs.

Lambing time—Season of the year when ewes normally bear their young.

Lanolin—Purified wool grease. It is used as a base for salves, ointments, in cosmetics, and for many other purposes.

Late lambs—Lambs born after the normal lambing time for a particular area has passed.

Marking—A term applied to docking and castrating lambs; also branding or marking for identification.

GLOSSARY OF COMMON TERMS

Mohair—Fleece of the Angora goat.

Mutton—The meat of sheep too mature to sell as lamb. Usually older than 14 months.

Open-faced—Sheep with little or no wool on the face, especially around the eyes.

Orphan—A lamb who has lost its mother.

Overshot jaws—A condition where the upper jaw is longer than the lower jaw, preventing the teeth from meshing properly. The opposite condition is known as **undershot**.

Parasites—(See Chapter VIII.)

Parrot mouth—A peculiar condition in the shape of the mouth resulting from one jaw crossing over the other.

Pinning—Collection of dung around the vent of very young lambs that has dried to the point of interfering with normal bowel movements.

Polled—Naturally hornless.

Pulled wood—Wool removed from the pelts of slaughtered lambs and sheep (occasionally from those that have died and in which decomposition is occurring).

Purebred—An animal of purebreeding, eligible for registration in the breed association.

Purity (wool)—Absence of fibers other than wool.

Rack—The area of the body, or especially the carcass, containing the ribs but excluding shoulder, breast, and loin.

Ram (buck)—An uncastrated male sheep of any age.

Saddle—Area in back of the shoulders.

Scoured (scouring)—Washed to remove yolk, dirt, and other natural impurities found in wool.

Second cuts—Short bits of wool resulting from passing shears twice over the same area in an effort to get close to the skin.

Set-up—Act of turning a sheep on its side or into a sitting position as a method of restraining.

Shear—The act of removing the fleece from the body of the sheep.

Shorn—Refers to sheep after the fleece has been removed.

Shrinkage—Percentage of the weight of grease wool lost in scouring.

Soundness

    Animal—Freedom from body defects.

    Wool—Strength of fiber.

Spinning count—Arbitrary numbers such as 40's, 50's, 70's denoting the degree of fineness of the fiber.

Stained—Wool colored from contact with urine or manure, or by bacterial action.

Staple—A lock or small sample of wool from a fleece.

Stud ram—A ram used as a sire in a purebred, registered flock.

Tags—Heavy, dungy wool.

Trimming—The act of removing part of the wool from a sheep in order to improve its appearance or facilitate normal functioning of the body.

Vaccinate—Introduction of antibiotics into animals so as to produce an immunity or tolerance to a disease.

GLOSSARY OF COMMON TERMS

Virgin wool—Generally means wool not previously used in the manufacture of fabrics, although the definition varies in different states.

Wether—A male sheep that was castrated at any early age, usually within two weeks after birth. Wether lambs, yearling wethers, etc.

Woolen—Fabrics or yarns made of uncombed wool.

Wool type—A, B, or C type wool refers to the wrinkle or fold development of the skin.

  A type—Has the greatest development. This type sheep has wrinkles more or less over the entire body. Generally of Merino blood.

  B type—Folds in this type are on neck and shoulders, and generally such sheep are superior in mutton quality to A type.

  C type—Few wrinkles on neck; none on body region. This type has the least fold development.

Worsted—Fabrics or yarns made of combed wool.

Yearlings—Young sheep that are approximately one to two years of age. They are identified by the fact that they have cut their first pair of permanent teeth but not the second pair.

Yield

  Carcass—Percentage of the live carcass left after removing entrails. Head and pelt may or may not be included, depending upon style of dressing.

  Wool—Percentage of the weight of grease wool left after scouring.

Yolk—The natural grease and suint covering of the wool fibers of the unscoured fleece, and excreted from glands in the skin.

INDEX

A

Aborted lamb, 169
Adaptability of sheep, 34
Advertising, 343
Africa, 4
Age of sheep, 128, 374
Analyzing records, 357
Antibiotics, 194
Approved practices, 89, 322
Arrangement of buildings, 213
Auctioneer, 343
Australia, 2
Authorities, 171, 334

B

Barbados, 23
Basque, 1
Beet pulp, 191
Bill of sale, 354
Birth, 146, 147
Blackbuck, 22
Blanketing, 134
Bloat, 252
Blocking, 134
Bluetongue, 286
Branding, 127, 157
Break joint, 294, 295, 303
Breaks, 109
Bred ewes, 140
Breeding, 70, 75
Breeding ewe, 58
Breeding ram, 61
Breeds, 37
Brooder, 154, 226
Burdizzo, 117
Butchering, 291
Buying a ram, 63
By-products, 199

C

Canning lamb, 310
Carcass, 298
Carding, 339
Castration, 118
Catching sheep, 114
Characteristics of breeds, 39
Chemical poisoning, 283
Chemical shearing, frontispiece, 102
Cheviot, 39
Chilled lamb, 152
Closebreeding, 73
Columbia, 40
Commercial feeder, 14
Commercial flock, 6
Concentrates, 182
Condition, 74
Contract, 201
Cooking, 304
Cooking timetable, 306
Corrals, 220
Corriedale, 40, 41
Coyotes, 170, 250
Credit, 359
Creep, 225
Creep feeding, 161, 204

Creep rations, 193
Crisscrossing, 72
Crossbreed, 71
Cull, 60, 67
Culling, 78
Curing lamb, 311
Cutting chute, 224, 225
Cyclophosphamide, frontispiece

D

Deficiencies, 189
Defleecing, frontispiece
Diagnosing pregnancy, 135
Dipping, 133
Dips, 229, 230
Direct crossing, 71
Disease, 245
Diseases of lambs, 163
Disguising odor, 153
Disinfecting, 15
Docking, 118
Dodge gate, 235
Dogs, 115, 171, 249, 251
Dorset, 18, 42
Dosing, 261
Drench, 133
Drenching, 259
Dual-purpose, 7
Dyeing sheep, 324

E

Ear notch, 166, 167
Ecology, 25
Elastrator, 118
Emergency feed, 112
Entry blank, 356
Equipment, 207
Everted womb, 159
Examining fleece, 107, 110
Exotic animals, 21

F

Farm flock, 8, 209
Federal lands, 11
Feeder, 189, 365
Feeder lamb, 56
Feeding for show, 202
Feeding guide, 187
Feed lot fattening, 15
Feed space requirement, 243

Fencing, 220
Feral, 22
Fiber grades, 326
Filing system, 353
Financial records, 349
Financial statement, 358
Fine wool, 37
Fitting, 337
Fleece, 57
Flocking instinct, 109
Floors, 98
Flushing, 74
Fly strike, 266
Foetus, 149
Foot rot, 270
Four-horned, 24
Foxhead, 158
Freezing, 308
Fur type, 38

G

Gestation table, 76
Glossary, 393
Good buys, 196
Grades, 64, 106
Grafting, 153, 368
Gregariousness, 109

H

Hampshire, 43
Handling, 93, 111, 332
Hay rack, 214, 215
Hazards, 6
Holding a sheep, 108, 332
Home consumption, 20
Home locker, 361
Hothouse lambs, 18
Hurdle, 218

I

Identification, 79
Identification tag, 373
Improving sheep, 69
Inbreeding, 73
Indians, 1
Insuring, 334
Inventory, 351
Irrigated pasture, 17

K

Killing, 292

INDEX

L

Ladino, 178
Lamb chart, 305
Lambing paralysis, 145
Lambs, 139
Landrace, 26, 44, 45
Liberia, 23
Life cycle, 256, 257
Lincoln, 47, 48
Linebreeding, 73
Litters of lambs, 45
Loading chute, 234

M

Malnutrition, 175
Marketing, 315
Marketing wool, 325
Market lamb, 15
Marking, 116
Mating, 77
Meat cuts, 305
Merino, 46
Milk fever, 289
Minerals, 187
Miscellaneous equipment, 232
Miss Wool, 316
Mouflon, 23
Mouth, 59
Mutton type, 38

N

Native sheep, 4
Navaho ram, 24
Navel, 151
Newborn, 151
New Zealand, 22
Nomadic, 11
Numbering, 80
Number per car, 333

O

Openface, 88
Opportunities, 8
Orphan, 21, 155, 156, 367
Outcrossing, 72
Overcrowding, 336
Oxford, 46

P

Painting ram, 127

Panels, 148
Parasites, 245
Partitions, 210
Parts of a sheep, 395
Parturition, 144
Pastures, 16, 176
Pelleting, 205
Pelt, 294
Pen layout, 95
Percentage lamb crop, 69, 73
Perspective, 30
Photographing, 344
Pivot gate, 221
Pocket fieldbook, 353
Poisonous plants, 280
Population, 6
Predators, 5, 169, 249
Preparing wool, 321
Prevention of illness, 246
Processing wool, 366
Progeny testing, 83
Prolapse, 168
Protection, 211
Purchase contract, 329
Purebred, 13

Q

Quadruplets, 27
Quality lambs, 56

R

Racks, 208
Ram, 77
Rambouillet, 49
Range band, 10
Range sheep, 185
Rations, 188
Records, 84, 86, 87, 347
Refrigeration, 364
Replacements, 81
Research, 27
Retail cuts, 299
Ring latch, 222
Rolled-in eyelid, 164
Romney, 38
Rotate, 77
Rotation, 181
Rotation crossing, 72

S

Sacking, 105
Sale, 64, 336, 341, 342

Salt, 189
Salt box, 260
Sanitation, 246
Scales, 228
Score card, 66
Selecting lamb and mutton, 301
Selling, 336
Setting up, 113
Shade, 181
Shearing, 94, 96, 101
Sheds, 96, 98
Sheep coat, 337, 338
Sheeping down, 198
Shipping, 331
Shipping crate, 237
Shipping fever, 269
Show date, 65
Show ring, 342
Show sheep, 131
Shrink, 58
Shropshire, 50, 70
Silo, 239
Skills, 371
Smoking lamb, 312
Sore mouth, 278
Southdown, 50
Space requirement, 227
Specialty producer, 17
Spinning count, 107
Sportsman, 173
Spring lamb, 19
Squeeze, 229
Standards, 80
Staple, 221
Stiff lamb disease, 165
Storage, 237
Straw, 346
Suffolk, 13, 53, 54
Summary, approved practices, 385
Summary, approved wool practices, 322
Summary, national program, 28
Supplement, 179
Synthetics, 3

T

Tables, 381
Tagging, 121, 141, 372
Targhee, 55
Tattooing, 131
Teats, 60
Teeth, 129
Toxicity, 283
Trends, 25
Trimming, 340
Trimming feet, 123, 124, 125
Trough, 216, 217
Twine, 105
Twins, 155
Tying fleece, 104
Types, 36

U

Udder, 60, 143
Unloading, 335
Urea, 200

V

Vaccination, 272
Ventilation, 209
Viscera, 296
Vitamins, 187

W

Waste feeds, 364
Wasteland, 4
Water, 191
Watering facilities, 219
Weaning, 167
Winter rations, 192
Wool, 4
Wool blindness, 52, 248
Wool packing rack, 241
Wrapping, 308